Speciality Chemicals in Mineral Processing

Speciality Chemicals in Mineral Processing

Edited by

D.R. Skuse
IMERYS Minerals, Sandersville, GA, USA

ROYAL SOCIETY OF CHEMISTRY

The proceedings of the meeting 'Speciality Chemicals in Mineral Processing' held at the University of Bath on 25–26 June 2001.

Special Publication No. 282

ISBN 0-85404-831-6

A catalogue record for this book is available from the British Library

Published by The Royal Society of Chemistry,
Thomas Graham House, Science Park, Milton Road,
Cambridge CB4 0WF, UK

Registered Charity Number 207890

For further information see our web site at www.rsc.org

Typeset in Great Britain by Vision Typesetting, Manchester
Printed by MPG Books Ltd, Bodmin, Cornwall, UK

Preface

This review comprises most of the presentations given at the RSC Industrial Affairs Division Speciality Chemicals Sector conference titled 'Speciality Chemicals in Mineral Processing' held in Bath, UK in 2001. The papers presented here are not intended to represent a comprehensive review of the use of speciality chemicals in mineral processing but rather the intention is to highlight a significant advance in chemical use in each of a number of different mineral process unit operations. The book is divided into three sections; (i) dispersion and flocculation, (ii) selective processing and (iii) microbiological control. Within each of these areas papers are presented that cover some of the most important recent research and developments from both academia and industry.

The first six papers are devoted to dispersion and flocculation. The plenary paper by Hogg reviews 'The Role of Polymers in Dispersion Stability' from both a theoretical and practical standpoint and sets the stage for the remaining chapters in the section. In the second chapter, Armes introduces the latest thinking from academia with respect to synthesis of surfactants and dispersants. The ambient temperature atom transfer radical polymerisation techniques described offer considerable opportunities for the future. The third chapter by Heyes reviews recent progress in using molecular simulation to understand the colloidal interactions of concentrated slurries and their role in determining slurry rheological behaviour. The chapters by Duccini and Gole present newly available dispersants and exemplify their behaviour in a number of mineral applications. Finally, the chapter by Phipps presents a colloid chemical analysis of the mechanism by which many polyelectrolyte dispersants work and makes suggestions as to how to improve dispersant performance.

There are four chapters on selective processing. The first of these chapters, by Dalton, focuses on using XPS and AFM techniques for the direct study of selective reagent adsorption. Next, Breen presents recent developments in using variable temperature DRIFTS to characterise in detail the interactions of re-agents with minerals. This is followed by a chapter by Warne which updates progress on recent work using molecular simulation to construct mineral surfa-ces and model their interactions with solvents and reagents. The final chapter of the section is by Gorken who describes the development and implementation of hydroxamate-based improved kaolin flotation reagents.

The final section of the book is devoted to microbiological control of mineral slurries. The first paper by Duddridge explores the EC's recent Biological Products Directive and speculates on the challenges that this will create for biocide producers, formulators and mineral slurry processors. The final two papers by Gutherie and Martin both introduce new microbiological control regimes.

It is a great pleasure to acknowledge all the contributors to this volume, without whom the conference and this book would not have been possible. In addition, it is a pleasure to acknowledge the support of the exhibitors who contributed to the success of the conference from which this book is derived, namely; Rohm and Haas, Ondeo-Nalco, Quantachrome, and Scientific and Medical products Ltd. I am grateful to Elaine Wellingham of the RSC for administrative help and to Dr Susan Partridge for considerable help with the manuscript. Finally and particularly, acknowledgement must go to my employer IMERYS Minerals (formerly English China Clays) for supporting this endeav-our and for direct financial support for four of the projects reported in this volume.

Contents

Microbiological Control

Dispersion and Flocculation

The Role of Polymers in Dispersion Stability

R. Hogg and C. Rattanakawin

DEPARTMENT OF ENERGY AND GEO-ENVIRONMENTAL
ENGINEERING, THE PENNSYLVANIA STATE UNIVERSITY,
UNIVERSITY PARK, PA 16802, USA

1 Introduction

Soluble polymers are widely used to control the state of dispersion of fine-particle suspensions. Depending on the polymer, and how it is applied, they can serve to enhance stability (dispersants) or to promote aggregation of the particles (flocculants). The topics covered in this chapter are intended as an overview of the use of polymers for stability control in mineral-particle suspensions with particular emphasis on flocculation processes. A brief discussion of stabilisation by polymers is included for completeness.

1.1 Polymeric Dispersants

Both natural and synthetic polymers with molecular weight up to about 20 000 are commonly used for stabilisation of fine-particle dispersions. It is generally accepted that these reagents function by presenting a steric barrier to direct contact between particles (steric stabilisation).[1] In effect, adsorption of the soluble polymer provides a lyophilic film on the particle surfaces that prefers contact with the solvent to contact with a similar film on another particle. The basic requirement for steric stabilisation is that the adsorbed layer be thick enough to prevent particles from approaching each other closely enough for attractive forces (*e.g.* van der Waals) to cause adhesion. More or less complete coverage of the surfaces is generally necessary. Charged polymers (polyelectrolytes) can further enhance the effect by adding an electrostatic repulsion (electrosteric stabilisation). Addition of the electrostatic component may serve to relax the coverage requirements for purely steric stabilisation.

1.2 Polymeric Flocculants

Polymers used to promote flocculation normally have higher molecular weight than the dispersants – usually from about 50 thousand up to about 20 million. Most of those in use today are synthetic, linear-chain molecules, and include non-ionic, anionic and cationic types. Non-ionic flocculants are commonly polyacrylamide, while the anionic reagents are often acrylamide/acrylate copolymers. Typical non-ionic and anionic flocculants have high molecular weights – in the 10 to 20 million range. Cationic flocculants (sometimes referred to as coagulants) are normally of lower molecular weight (<1 million) and commonly owe their charge to the presence of a quaternary ammonium group. The polydiallyldimethylammonium chloride (DMDAAC) types are widely used commercially. Charge densities are usually substantially higher for the cationic reagents than for the anionic types.

   Flocculation by polymers, like dispersion, normally involves adsorption at particle surfaces. The high molecular weight flocculants are generally agreed to function through a 'bridging' mechanism whereby a single large molecule can attach to two or more particles simultaneously, providing a physical link – bridge – between them.[2-4] Smaller, highly charged polyions can adsorb on oppositely charged particle surfaces forming patches of opposite charge. Thus, a cationic polymer can form positively charged patches on the surface of negatively charged particles. Interaction of a patch on one particle with a region of bare surface on another then leads to aggregation.[3] A third mechanism, known as depletion flocculation, does not involve adsorption. Rather, the exclusion of large polymer molecules from small spaces between particles in concentrated suspensions promotes aggregation through osmotic pressure effects.[4]

1.3 Polymer Solutions

The properties of polymer solutions play an important role in the dispersion or flocculation of fine-particle suspensions. Individual molecules in solution generally take on a more-or-less randomly coiled configuration.[5] Repulsion between similarly charged ionic groups in a polyelectrolyte leads to expansion of the coil, increasing the effective size of the molecule in solution. Modification of the charge, *e.g.* by protonation of acid groups in an anionic polymer, gives rise to changes in molecular conformation. Shielding of the charges in the presence of simple electrolytes generally leads to increased coiling, *i.e.* smaller effective molecular size.

2 Polymer Adsorption

The adsorption of polymers at solid surfaces is substantially more complicated than that of small molecules.[6,7] Typically, a small molecule adsorbs by attachment of a functional group on the molecule to a site on the surface. For the case of adsorption from aqueous solution, the process involves displacement of water from sites at the surface. The extent of adsorption is determined by the preference of the surface site and the adsorbing species for contact with water rather than with each other. While the same is generally true for polymers, the presence of a large number of potential attaching groups on each molecule biases the system towards adsorption. A typical polyacrylamide flocculant with a molecular weight of 10 million has about 140 000 individual segments, each of which is capable of attaching to the surface. Only one such attachment is necessary for adsorption of the whole molecule. The consequences of this 'multiple adsorption' effect are:

- Polymer adsorption tends to be indiscriminate. The large molecules become attached to essentially any available surface unless strongly repelled from it.
- Adsorption is effectively irreversible. Desorption of the molecule requires simultaneous detachment of all adsorbed segments.
- Since the adsorption of individual segments is quite reversible, the conformation of the adsorbed molecules can vary continuously and the molecules are free to migrate over the solid surface.
- The strength of the segment–surface site interaction has more effect on the conformation of the adsorbed molecules than on the number of molecules adsorbed.
- The effective 'parking area' of the adsorbed molecule may vary with the extent of adsorption. As more molecules are adsorbed, the area occupied by each may decrease due to a crowding effect.

When the adsorbing polymer acts as a flocculant, the adsorption process is further complicated by concurrent aggregation of the particles. Adsorption of the polymer provides the linkages needed for bridging flocculation while the formation and growth of aggregates reduces the surface area available for adsorption. An interesting consequence is that effective flocculation actually reduces the amount of polymer adsorbed.[8,9] An example of this inverse relationship is given in Figure 1. Adsorption densities at pH 4.5 and 11, where the dispersion was initially stable and flocculation by the polymer was poor, are more than ten times greater than those on the flocculated suspension at pH 9.

Because polymer adsorption is effectively irreversible, and because adsorption and floc growth occur simultaneously, flocculation is a non-equilibrium process. As a result, performance is largely determined by the kinetics of adsorption and aggregation. Both of these can be regarded as collision processes involving solid particles and polymer molecules. In each case, collisions can arise due to either Brownian motion or agitation of the suspension. The collision frequency v between particles and polymer molecules can be estimated from:[10]

$$v = K n_p n \tag{1}$$

where n and n_p are the respective number concentrations for solid particles and polymer molecules and K is a rate constant. For Brownian motion:

$$K = \frac{2kT}{3\mu}\left(1 + \frac{x_p}{x} + \frac{x}{x_p}\right) \tag{2}$$

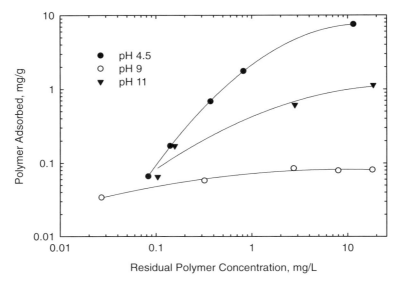

Figure 1 *Effect of pH on the adsorption of a non-ionic polyacrylamide on fine alumina particles (after Ray)*[10]

where k is Boltzmann's constant, T is absolute temperature, μ is the fluid viscosity and x and x_p are the sizes of the particles and molecules in solution. In the case of collisions due to agitation:

$$K = \frac{G}{6}(x + x_p)^3 \tag{3}$$

where G is the mean shear rate due to agitation.

The relative importance of the two mechanisms can be seen in Figure 2. The curves indicate, for different shear rates, molecular weight/particle size combinations for which the rates are equal. Combinations to the right of the lines represent domination by shear. It is clear that the rates are dominated by the shear mechanism for most conditions of practical interest.

3 Polymers and Dispersion Stability

Dispersions of fine mineral particles can be stabilised by direct electrical charging of the particles or by steric/electrosteric protection from adsorbed polymers. Stabilisation by direct charging is well described by the classical DLVO theory.[11]

3.1 Steric Stabilisation

Dispersion stabilisation generally requires fairly complete coverage of particle

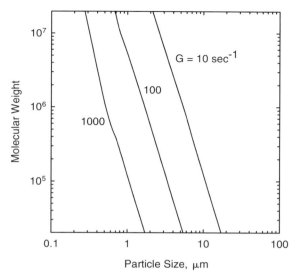

Figure 2 *Relative adsorption rates due to collisions resulting from Brownian motion and mechanical agitation. The lines represent conditions where the rates are equal. Agitation dominates to the right of the lines*

surfaces by adsorbed polymer. Polymer molecular weight should be sufficiently large to prevent close approach of the particles, but not so large as to promote bridging flocculation. Purely steric stabilisation usually involves high polymer dosage – of the order of 1 mg per square metre of solid surface.

3.2 Electrosteric Stabilisation

In many practical cases, stabilisation by polymers involves a combination of steric and charge interactions. Unlike simple electrolytes, multiple adsorption effects permit polyelectrolytes to continue to adsorb well beyond the point where the adsorbed layer charge exceeds that of the particle surface. In this way, the effective charge on particles can be increased substantially at relatively low surface coverage by the polymer.

4 Flocculation by Polymers

In general, flocculation processes involve two basic steps:

• Destabilisation of the dispersion to permit particle–particle collisions and aggregate formation.
• Growth of small aggregates to form large flocs.

Polymers can play a role in both of these.

4.1 Destabilisation

The stability of suspensions of fine mineral-particles usually results from the electrical repulsion forces that prevent association of particles into aggregates. These can be minimised by reducing surface potentials – often by appropriate pH adjustment – or by shielding of the charge using simple electrolytes. In principle, polyelectrolytes can also serve to shield the charge on particle surfaces, but the charge-patch and bridging mechanisms described above are probably more important in practice.

An example of destabilisation by a polyelectrolyte, probably through the charge-patch mechanism, is given in Figure 3. The figure shows the evolution of the floc size distribution with the addition of a cationic polymer (molecular weight about 100 000) to a suspension of fine alumina particles in water at pH 11. Initially the dispersion was stable due to the negative charge on the particles at that pH. Upon addition of the polymer, the distribution first becomes bimodal, consisting of small flocs and some residual primary particles. As more polymer is

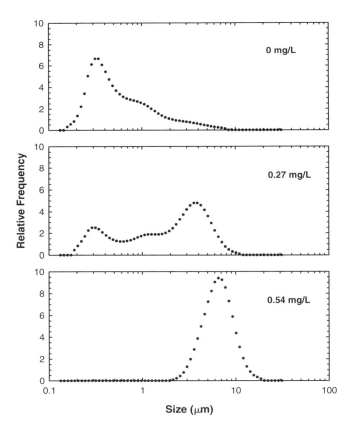

Figure 3 *Destabilisation of a fine alumina suspension (3% solids by weight) at pH 11 by a cationic polyelectrolyte (molecular weight about 100 000)*

added, the primary-particle mode disappears and the distribution becomes unimodal again at a polymer concentration of about $0.5\,mg\,L^{-1}$ ($0.0167\,mg$ per gram of solids). However, further polymer addition has very little effect, other than increasing the median floc size from about $6–7\,\mu m$ at $0.5\,mg\,L^{-1}$ to about $9\,\mu m$ at $5\,mg\,L^{-1}$. Apparently, the cationic polymer is effective for destabilising the suspension but not for promoting floc growth.[12,13]

Very similar effects can be seen for higher molecular weight (about 15 million) non-ionic and anionic polymers. Some results for the addition of a non-ionic polyacrylamide to the same alumina under the same conditions are given in Figure 4. The unimodal–bimodal–unimodal progression can be seen again. The principal differences from the cationic polymer are that more polymer is needed to eliminate the primary-particle mode (about $3\,mg\,L^{-1}$) but continued polymer addition leads to significant growth of the flocs – up to about $80\,\mu m$ at $5\,mg\,L^{-1}$ of the polymer. For the anionic polymer, even more polymer is required to eliminate the primary particles ($10\,mg\,L^{-1}$) and floc growth continues up to more than $200\,\mu m$. It seems that the higher molecular weight polymers are less

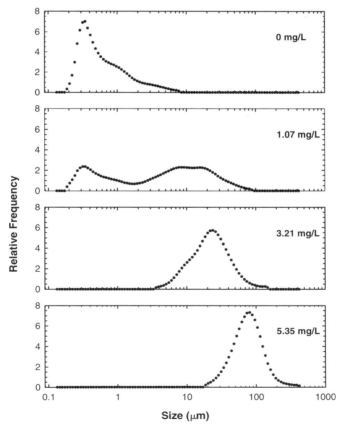

Figure 4 *Destabilisation and flocculation of a fine alumina suspension (3% solids by weight) at pH 11 by a non-ionic polyelectrolyte (molecular weight about 15 million)*

effective for destabilisation of the primary particles but more effective for promoting floc growth. The anionic polymer is able to induce flocculation despite the similar charge on the polymer and the particles. Apparently, the high molecular weight permits adsorption and bridging to occur.

It is interesting to compare destabilisation by polymers with equivalent results for simple electrolytes. The effects of adding calcium chloride to the alumina suspension at pH 11 are shown in Figure 5. At a salt concentration of about 10^{-4} moles L^{-1} (11 mg L^{-1}), the size distribution changes abruptly from that of the primary particles to a unimodal floc size distribution centred at about 6 μm.

Destabilisation of the alumina at pH 5 where the particles are positively charged shows quite similar trends, the principal difference being that the cationic polymer is completely ineffective.[15] Presumably, the positive charges on both particles and polymer are sufficient to prevent adsorption of the relatively small, highly charged polyions. The behaviour of the non-ionic and anionic polymers is essentially the same as at pH 11.[15] Destabilisation by the addition of

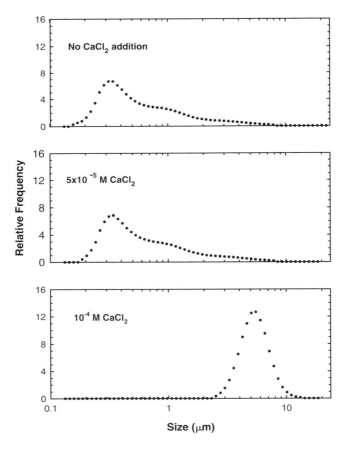

Figure 5 *Destabilisation of a fine alumina suspension (3% solids by weight) at pH 11 by the addition of a divalent cation*

divalent sulfate ions at pH 5 is virtually identical to that observed for calcium at pH 11.[15] Adjustment of the pH to around the point of zero charge (pH 9) leads to the same floc size distributions as the simple salt additions.[15]

4.2 Floc Growth

Collisions and adhesion between particles/flocs lead to floc growth. The collision mechanisms described previously for polymer adsorption apply equally to floc formation and growth. Collisions due to shear can be expected to dominate for most practical conditions. In order for growth to continue, it is necessary for the collisions to lead to adhesion and for the resulting flocs to be resistant to degradation and breakage in the agitated system. Destabilisation of the dispersion should ensure that particles adhere on contact, but the forces involved will not necessarily be sufficient to withstand disruption of a large floc subjected to shear. In processes for deaggregation or breakage of particles under applied stress, it is generally observed that breakage rates increase with increasing particle or aggregate size. It follows that simultaneous growth and breakage in flocculation processes can be expected to lead to the approach to a limiting floc size, at which growth and breakage rates are equal. Stronger flocs should grow to larger limiting sizes.

It is postulated that polymers can perform a dual function in flocculation. They can aid in destabilisation as noted above and also serve as binders in the growing flocs. The results shown in Figures 3–5 are clearly consistent with this concept. Destabilisation by pH control or simple salt addition is sufficient to allow individual particles to adhere on contact, but the resulting flocs are weak and can only grow up to a relatively small limiting size (about $6\,\mu$m for the conditions applying in Figure 5). The use of a polymer produces stronger flocs as reflected in the larger limiting sizes shown in Figures 3 and 4: over $100\,\mu$m for the high molecular weight non-ionic and anionic polymers. The low molecular weight cationic polymer appears to provide only a minor increase in floc strength.

It is apparent that pH control or the addition of simple salts or low molecular weight polyelectrolytes is typically sufficient for destabilising fine-particle dispersions, while high molecular weight polymers are necessary to permit floc growth to large sizes. It follows that the use of reagent combinations should offer considerable potential for efficient flocculation. Some examples of this approach are illustrated in Figures 6 and 7. The effects of pH control and salt addition can be seen in Figure 6. Addition of a non-ionic polymer directly to the stable dispersion requires $9-10\,\text{mg L}^{-1}$ to produce flocs with a median size of about $50\,\mu$m. By first adding $10^{-2}\,\text{moles L}^{-1}$ of sulfate ions or adjusting the pH to about 8, the same floc size can be obtained with a polymer addition of less than $2\,\text{mg L}^{-1}$.

The use of polymer combinations is demonstrated in Figure 7. The addition of about $1\,\text{mg L}^{-1}$ of a low molecular weight cationic polymer to negatively charged alumina particles at pH 11 destabilises the dispersion but produces only very small flocs – around $7\,\mu$m (Figure 7a). Further treatment, with about

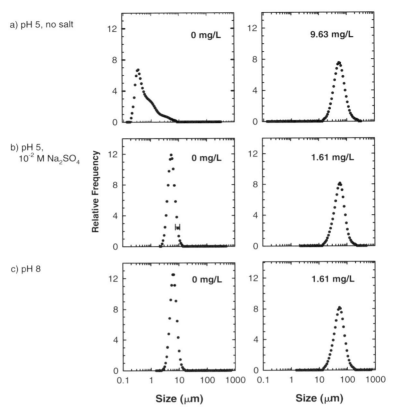

Figure 6 *Effect of destabilisation by salt addition or pH control on flocculation with a non-ionic polymer*

$2\,\mathrm{mg\,L^{-1}}$ of a high molecular weight anionic polymer, causes substantial floc growth – to over $100\,\mu\mathrm{m}$ (Figure 7b).

In the use of these reagent combinations, where destabilisation and floc growth are addressed separately, the order of treatment is extremely important. Destabilisation must be achieved first, before the growth-enhancing polymer is applied. An example of the effect of reversing the sequence can also be seen in Figure 7. Addition of about $2\,\mathrm{mg\,L^{-1}}$ of the anionic polymer to the stable dispersion produces a bimodal floc size distribution consisting of a mixture of flocs at around $30\,\mu\mathrm{m}$ and primary particles at around $0.5\,\mu\mathrm{m}$ (Figure 7c). The addition of $1\,\mathrm{mg\,L^{-1}}$ of the cationic polymer causes aggregation of the residual primary particles, shifting the fine-particle mode to about $10\,\mu\mathrm{m}$ but has only minor effects on the coarser mode (Figure 7d). If anything, addition of the cationic polymer seems to reduce the height of the coarse mode, implying a transfer of particles from the coarse to the fine modes.

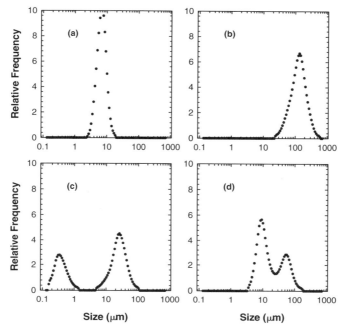

Figure 7 *Destabilisation/flocculation of a fine alumina suspension at pH 11 by polymer combinations. (a) Cationic only ($1.07\,mg\,L^{-1}$); (b) cationic ($1.07\,mg\,L^{-1}$) followed by anionic ($2.14\,mg\,L^{-1}$); (c) anionic only ($2.14\,mg\,L^{-1}$); (d) anionic ($2.14\,mg\,L^{-1}$) followed by cationic ($1.07\,mg\,L^{-1}$)*

5 The Flocculation Process

5.1 Mixing

Chemical treatments to induce flocculation in fine-particle dispersions are generally sensitive to the process of adding reagents.[4,14,15] Mixing of reagents with the suspension is usually accomplished by some kind of mechanical agitation. It is important to recognise, however, that agitation also promotes both growth and breakage of flocs. As noted above, the balance between growth and breakage rates typically leads to the approach to a limiting floc size. For systems involving only simple reagents such as inorganic salts, chemical equilibrium is probably reached also. On the other hand, the effective irreversibility of polymer adsorption makes the establishment of equilibrium unlikely for cases of polymer-induced flocculation. In fact, it can be demonstrated that the approach to equilibrium is usually detrimental to the process and that efficient flocculation involves exploiting the non-equilibrium nature of the process.[16]

Because polymer adsorption tends to be irreversible, inadequate mixing, with non-uniform distribution of polymer in the particle suspension, cannot be rectified by subsequent rearrangement. Locally high concentrations can lead to rapid floc growth, but with corresponding regions of starvation elsewhere in the system. The problem is especially acute at high solids concentrations because of the very high rates of adsorption. It follows that the time required for polymer adsorption should be greater than the 'mixing time'. For stirred tanks, the mixing time t_m can be estimated from:[17]

$$t_m \approx 36/N \tag{4}$$

where N is the rotational speed of the impeller. Thus, for the mixing system used in the experiments described above ($N = 1000$ rpm), mixing times of around 2 seconds can be expected.

The corresponding adsorption time can be estimated using Equations 1 and 3. The number concentration of particles is given by:

$$n = \frac{\phi}{k_v x^3} \tag{5}$$

where ϕ is the volume fraction of solids in suspension and k_v is the volume shape factor. Neglecting the molecular size x_p in Equation 3, the adsorption rate (assumed to be equal to the collision frequency) can be estimated from:

$$\frac{dC_a}{dt} = \frac{\bar{G}\phi}{6k_v} C_p \tag{6}$$

where C_a and C_p are respectively the concentrations of adsorbed and free polymer in the system. If the overall polymer dosage is C_{po}, added at time zero:

$$C_p = C_{po} - C_a \tag{7}$$

and the solution to Equation 6 is:

$$C_a = C_{po}\left[1 - \exp\left(-\frac{G\phi}{6k_v}t\right)\right] \tag{8}$$

For the results shown in Figures 3–6, $G \approx 1000\ \text{sec}^{-1}$ and $\phi \approx 0.01$. Assuming roughly spherical particles, $k_v \approx \pi/6$. The time needed for 90% of the polymer to be adsorbed would be about 3 seconds, which is quite similar to the mixing time. In other words, a significant fraction of the added polymer could be adsorbed locally before the mixing action could provide a uniform distribution. In order to minimise the detrimental effects of localised adsorption, it is generally preferable to add the polymer continuously over an extended period, substantially greater than the mixing time.

Since the same mechanisms are involved, aggregation times are generally similar to adsorption times. As a consequence, floc growth rates typically parallel adsorption rates or, for continuous polymer addition, the rate of polymer addition. Furthermore, floc growth continues only so long as fresh polymer is available in solution. Prolonged agitation, in the absence of fresh polymer leads to degradation of the flocs.[17,18]

5.2 Solids Concentration

The concentration of suspended solids is an important variable in flocculation and can vary from parts per million in potable water treatment to more than 20% by volume in filtration applications. Differences in solids concentration affect both the amount of polymer needed and the rates of adsorption and floc growth. Dosage requirements can reasonably be expected to vary in direct proportion to solids concentration but the constraints on the mode of polymer addition appear to be less simple. At high solids concentrations, the potential floc growth rate will invariably be higher than any reasonable rate of continuous polymer addition. Consequently, growth is controlled by the rate of addition. However, the limiting floc size is also affected by solids concentration. In addition to the breakage limitation described previously, floc size in very concentrated suspensions may be limited by a crowding effect. It is well known that floc structures become increasingly open as size increases. Thus, as flocs grow, they eventually approach a condition of close packing, thereby preventing further growth.

Somewhat different constraints apply at low solids concentrations. Adsorption and floc growth rates are much lower and mixing is a less critical factor. However, while growth rates are smaller, breakage rates are determined mostly by floc size and agitation intensity and should be relatively unaffected by solids concentration. The balance between growth and breakage can then be expected to occur at a smaller size. It follows that, at low solids concentration, it may be appropriate to employ lower agitation rates and allow an extended mixing period.

5.3 Polymer Solution Concentration

The concentration of the polymer solution added to the particle suspension has also been shown to affect process performance.[17] The use of more dilute solutions appears to enhance floc growth. In concentrated solutions, intermolecular repulsion enhances coiling of the molecules, reducing their effective size. Because of the very high adsorption rates, there is insufficient time for relaxation of the molecules, and the surface covered by a molecule and its extension into solution are both reduced.

6 Applications

The observations described above are based on small-scale laboratory testing. Application to industrial operations such as mineral processing introduces further questions. While chemical requirements can probably be evaluated directly from the laboratory tests, the physical aspects, particularly mixing and reagent addition, must be scaled-up appropriately. Furthermore, while laboratory testing is normally carried out in the batch mode, industrial processes are usually continuous.

6.1 Scale-up

Laboratory tests indicate that agitation intensity and the rate of polymer addition are important factors in determining process performance. Information on scale-up rules for these variables is, unfortunately, rather limited. Maffei[18,19] carried out an empirical study of the flocculation of kaolin with polyacrylamide in a series of geometrically similar stirred tanks ranging in volume from 0.3 to 52 litres. His results suggest the following rough guidelines for the scale-up of batch flocculation systems based on laboratory tests:

• Maintain the same overall polymer dosage and relative rate of addition (rate per unit volume). For continuous polymer addition this specifies the same mixing/reagent addition time.
• Scale the agitator speed to maintain the same impeller tip speed. Thus, for an impeller of diameter d:

$$N \propto 1/d \tag{9}$$

It should be emphasised that these are approximate guidelines only, based on quite limited experimental data.

6.2 Continuous Flocculation

Most industrial flocculation is carried out as a continuous process, either in a stirred tank or, perhaps more commonly, by direct injection of a polymer solution into a flowing stream. Again, only rather limited information is available on the behaviour and performance of such systems. A study of continuous flocculation in a stirred tank[20,21] has revealed some similarities and some differences relative to the equivalent batch system. An important inherent difference is that, whereas polymer addition in a batch system leads to increasing dosage over the period of addition, the polymer content of a continuous system at steady state is constant with time. It has been postulated that this difference permits a somewhat closer approach to equilibrium in the continuous system. The result is that, while similar trends are observed, the effects of mixing/reagent-addition time are much less dramatic for the continuous case. A comparison of the effects of mixing time on batch and continuous flocculation is given in Figure 8. In general, continuous flocculation appears to produce smaller, slower-settling flocs than the batch systems.

Flocculation during flow through pipes has also received some attention.[22,23] To some extent, laboratory-scale test results parallel those obtained in stirred tanks. In principle, the addition of polymer at a single location in a pipe is roughly analogous to instantaneous addition followed by a period of continued

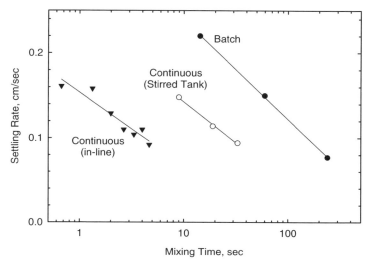

Figure 8 *A comparison of the effects of mixing time on batch and continuous flocculation of kaolin at 3% solids by weight by 4 mg L^{-1} of a non-ionic polymer (after Suharyono)*[22]

mixing in a stirred tank, operated in the batch mode. Similarly, addition at several points along the length of the pipe is equivalent to continuous addition to a tank. Experimental results are consistent with these similarities but, as for continuous operation in tanks, the flocs produced are generally smaller and exhibit lower settling rates than those formed in batch systems.[25]

Direct comparison between batch and continuous, in-line flocculation at the laboratory-scale is difficult. Since mixing action in pipes is provided by flow, mixing intensity and mixing time cannot be varied independently in a pipe of fixed length. Achieving high mixing intensities similar to those found to be appropriate in stirred tanks requires high flow rates through pipes of small diameter. Extremely long pipes are then needed in order to obtain similar mixing times. The same problem of matching mixing times and agitation intensities presents serious difficulties in scaling up to industrial conditions.

7 Conclusions

Polymers can be used either to stabilise fine-particle dispersions or to promote flocculation. Adsorption of the polymer at particle-solution interfaces is usually involved in both cases. The principal differences between the two are the molecular weight and relative amounts of polymer needed: high surface coverage by a low molecular weight polymer leads to stabilisation while flocculation requires much lower dosage of high molecular weight polymers. In the case of dispersion stabilisation, adsorption occurs on individual particles, which remain as discrete entities. Consequently, the process can be conducted under near-equilibrium conditions and performance depends largely on chemical factors. Flocculation, on the other hand, involves polymer attachment to more than one particle, accompanied by aggregation of the particles. The process occurs under non-equilibrium conditions and performance is strongly affected by physical factors such as mechanical agitation.

In general, flocculation involves two stages – destabilisation of the suspension and the growth of flocs. Destabilisation can be achieved by:

* pH control
* appropriate salt addition
* the use of low molecular weight polyelectrolytes

High molecular weight polymers are not recommended for destabilisation but can be highly effective in promoting floc growth in already destabilised systems. The order of treatment – destabilisation followed by flocculant addition – is very important.

The performance of high molecular weight polymeric flocculants is heavily dependent on the physical conditions of their application. Recommended treatment conditions are:

* addition from a dilute polymer solution
* rapid addition with vigorous agitation (short mixing time)
* at low solids concentrations (< 1% by volume), a follow-up period of relatively gentle agitation is useful to allow for floc growth
* continued agitation of a concentrated suspension after flocculant addition serves mainly to degrade existing flocs and should be avoided

Scale-up considerations for batch flocculation in stirred tanks indicate that similar results are obtained with:

* the same rate of polymer addition per unit volume
* the same mixing/polymer-addition time
* the same impeller tip speed

Continuous flocculation in stirred tanks shows similar trends with regard to agitation/mixing conditions as for batch systems. The overall performance of continuous flocculation processes tends to be inferior to that which can be obtained in batch systems. This is attributed to a closer approach to equilibrium in the continuous case. In-line flocculation, by injection of polymer into turbulent flow in a pipe, is widely practised in industry and offers certain advantages. Multiple addition at several points along the length of a pipe is preferred to single-point addition.

Nomenclature

C_a	Concentration of adsorbed polymer	$M\,L^{-3}$
C_p	Concentration of polymer in solution	$M\,L^{-3}$
C_{po}	Initial polymer concentration	$M\,L^{-3}$
d	Impeller diameter	L
G	Mean shear rate	T^{-1}

k	Boltzmann's constant	$ML^2T^{-2}\,{}^\circ K^{-1}$
k_v	Volume shape factor	$-$
K	Collision rate constant	L^3T^{-1}
n	Number concentration of particles	L^{-3}
n_p	Number concentration of polymer molecules	L^{-3}
N	Impeller rotational speed	T^{-1}
t	Time	T
t_m	Mixing time	T
T	Absolute temperature	$^\circ K$
x	Particle size	L
x_p	Effective size of polymer molecule	L
ϕ	Solids concentration (volume fraction)	$-$
μ	Liquid viscosity	MLT^{-1}
ν	Collision frequency	$L^{-3}T^{-1}$

References

1. D.A. Napper, *Polymeric Stabilisation of Colloidal Dispersions*, Academic Press, New York, 1983.
2. A.S. Michaels, *Ind. Eng. Chem.*, 1954, **46**, 1485.
3. V.K. La Mer and T.W. Healy, *Rev. Pure Appl. Chem.*, 1963, **13**, 112.
4. J.A. Kitchener, *Br. Polymer J.*, 1972, **4**, 217.
5. J. Gregory, *J. Coll. Interface Sci.*, 1973, **42**, 448.
6. G.J. Fleer, J.M.H.M. Scheutjens and B. Vincent, 'The Stability of Dispersions of Hard Spherical Particles in the Presence of Nonadsorbing Polymer' in *Polymer Adsorption and Dispersion Stability*, E.D. Goddard and B. Vincent (eds.), American Chemical Society, Washington DC, 1984, ACS Symposium Series **240**, Chapter 16, pp. 245–263.
7. P.J. Flory and T.G. Fox Jr., *J. Am. Chem. Soc.*, 1951, **73**, 1904.
8. J.M.H.M. Scheutjens and G.J. Fleer, *J. Phys. Chem.*, 1979, **83**, 1619.
9. M.A. Cohen-Stuart, T. Cosgrove and B. Vincent, *Adv. Coll. Interface Sci.*, 1986, **24**, 143.
10. D.T. Ray, PhD Thesis, The Pennyslvania State University, 1988.
11. D.T. Ray and R. Hogg, 'Bonding of Ceramics Using Polymers at Low Concentration Levels' in *Innovations in Materials Processing Using Aqueous, Colloid & Surface Chemistry*, F.M. Doyle, S. Raghavan, P. Somasundaran and G.W. Warren (eds.), The Minerals, Metals & Materials Society, Warrendale PA, 1989, pp. 165–180.
12. R. Hogg, *Colloids and Surfaces A*, 1999, **146**, 253.
13. J.Th.G. Overbeek, in *Colloid Science*, H.R. Kruyt (ed.), Elsevier, Amsterdam, 1952, Vol. I, Chapters VI and VII, pp. 245–301.
14. C. Rattanakawin, MS Thesis, The Pennyslvania State University, 1998.
15. C. Rattanakawin and R. Hogg, *Colloids and Surfaces A*, 2001, **177**, 87.
16. R.O. Keys and R. Hogg, 'Mixing Problems in Polymer Flocculation' in *Water – 1978*, G.F. Bennett (ed.), AIChE Symposium Series, American Institute of Chemical Engineers, New York, 1979, Vol. 75, pp. 63–72.
17. R. Hogg, P. Bunnaul and H. Suharyono, *Miner. Metall. Process*, 1993, **10**, 81.
18. R. Hogg, R.C. Klimpel and D.T. Ray, *Miner. Metall. Process*, 1987, **4**, 108.
19. L.A. Cutter, *AIChE J.*, 1966, **12**, 35.
20. A.C. Maffei, MS Thesis, The Pennsylvania State University, 1989.

21. R. Hogg, A.C. Maffei and D.T. Ray, 'Modeling of Flocculation Processes for Dewatering System Control' in *Control '90 – Mineral and Metallurgical Processing*, R.K. Rajamani and J.A. Herbst (eds.), Society for Mining, Metallurgy and Exploration, Littleton CO, 1990, Chapter 4, pp. 29–34.
22. H. Suraryono, PhD Thesis, The Pennsylvania State University, 1996.
23. H. Suharyono and R. Hogg, *SME Preprint* No. 94-231, Society for Mining, Metallurgy and Exploration, Littleton CO, 1994.
24. J. Gregory and L. Xing, 'Influence of Rapid Mix Conditions on Flocculation by Polymers' in *Dispersion and Aggregation: Fundamentals and Applications*, B.M. Moudgil and P. Somasundaran (eds.), Engineering Foundation, New York, 1994, pp. 427–441.
25. H. Suharyono and R. Hogg, *Miner. Metall. Process*, 1996, **13**, 93.

Synthesis of New Polymeric Surfactants and Dispersants *Via* Atom Transfer Radical Polymerisation at Ambient Temperature

S.P. Armes, K.L. Robinson, S.Y. Liu, X.S. Wang, F.L.G. Malet and S.A. Furlong

SCHOOL OF CHEMISTRY, PHYSICS AND ENVIRONMENTAL SCIENCE, UNIVERSITY OF SUSSEX, FALMER, BRIGHTON, EAST SUSSEX, BN1 9QJ, UK

1 Introduction

Atom Transfer Radical Polymerisation (ATRP) was discovered independently by Wang and Matyjaszewski,[1] and Sawamoto's group[2] in 1995. Since then, this field has become a 'hot topic' in synthetic polymer chemistry, with over 1000 papers published worldwide and more than 100 patent applications filed to date. ATRP is based on Kharasch chemistry; overall it involves the insertion of vinyl monomers between the R–X bond of an alkyl halide-based initiator. At any given time in the reaction, most of the polymer chains are capped with halogen atoms (Cl or Br), and are therefore dormant and do not propagate; see Figure 1.

The role of the copper-based ATRP catalyst is to remove these terminal halogen atoms and generate polymer radicals. A short burst of propagation occurs, followed by rapid, reversible halogen capping of the growing polymer chain. Thus the instantaneous polymer radical concentration is significantly reduced relative to that normally found in conventional radical polymerisations. This means that the probability of any two polymer radicals terminating is suppressed relative to the probability of propagation (chain extension). Hence remarkably narrow molecular weight distributions can be achieved, in some cases as low as 1.05 for polystyrene.[3]

ATRP has been used to polymerise styrenic, acrylic and methacrylic monomers. However, conventional ATRP of styrenic and methacrylic monomers is rather sluggish: typically many hours are required at high temperatures ($> 90\,^\circ$C) for incomplete conversions ($< 90\%$) even under bulk polymerisation conditions. Acrylates polymerise much faster than methacrylates or styrenics and the rapid, efficient polymerisation of *n*-alkyl acrylates at room temperature with various

21

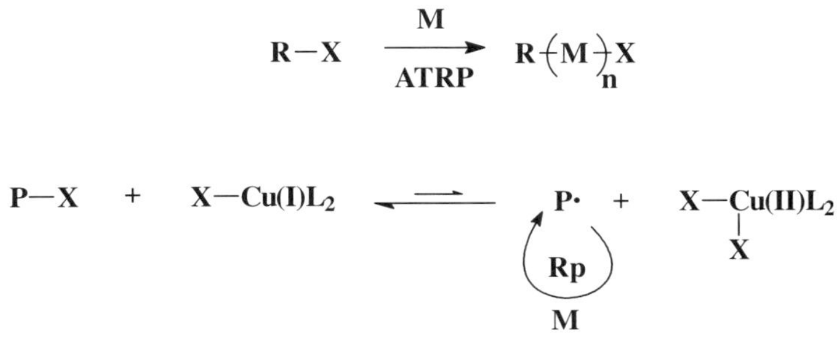

$$R-X \xrightarrow[\text{ATRP}]{M} R \text{-}(M\text{-})_n X$$

$$P-X \; + \; X-Cu(I)L_2 \; \underset{}{\overset{}{\rightleftharpoons}} \; P\cdot \; + \; X-Cu(II)L_2$$

Rp

M

P—X Dormant halide capped chain   X Halogen

P· Growing polymer radical   L Solubilising ligand

Figure 1 *General reaction scheme for Atom Transfer Radical Polymerisation (ATRP)*

multidendate ligands has been reported.[4]

Because ATRP is based on radical, rather than ionic, polymerisation chemistry, it has gained a deserved reputation for excellent tolerance towards protic solvents and impurities and ATRP has been exploited by many research groups to polymerise a wide range of functional monomers. Several years ago we discovered that ATRP of various hydrophilic methacrylates proceeds very rapidly in water under remarkably mild conditions. For example, mono-methoxy-capped oligo(ethylene glycol) methacrylate [OEGMA] was polymerised to more than 95% conversion within 25 minutes at 20 °C and GPC analysis indicated final polydispersities of 1.20 to 1.30. Similar results were obtained with sodium 4-vinylbenzoate [NaVBA] and 2-methacryloyloxyethyl phosphorylcholine [MPC]. Matyjaszewski and co-workers have previously reported a good correlation between the rate of ATRP and the solvent polarity, although the use of water was not investigated.[5] We have recently obtained ESR and UV–visible absorption spectroscopy data which indicate a somewhat higher Cu(II) concentration in the aqueous ATRP of OEGMA compared to bulk ATRP of the same monomer. This is consistent with the observed increase in rate of polymerisation in the former case and also accounts for the somewhat higher polydispersities, which result from the increased probability of termination.

If ATRP is conducted in methanol, under the same conditions, the rate of polymerisation is slower (95% conversion required 2–5 h at 20 °C). However, for several hydrophilic monomers such as 2-hydroxyethyl methacrylate [HEMA] and glycerol monomethacrylate [GMA], methanolic ATRP is the preferred method for optimal living character since narrower polydispersities (typically 1.10 to 1.20) and better blocking efficiencies are generally obtained.

Compared to other living radical polymerisation techniques, ATRP offers two important advantages. Firstly, the synthesis of well-defined macro-initiators is facile and allows the preparation of a range of new diblock copolymers. Secondly, the presence of water (or methanol) has a remarkable accelerating effect on the rate of ATRP of hydrophilic monomers, allowing high conversions to be achieved rapidly even under mild conditions.[6]

Herein we summarise our recent progress in the exploitation of ATRP for the synthesis of controlled-structure block copolymer surfactants and dispersants.

2 Experimental

We have recently evaluated the ATRP of a wide range of hydrophilic monomers such as 2-sulfatoethyl methacrylate (SEM), sodium 4-vinylbenzoate (NaVBA), sodium methacrylate (NaMAA), 2-(dimethylamino)ethyl methacrylate (DMA), 2-(*N*-morpholino)ethyl methacrylate (MEMA), 2-(diethylamino)ethyl methacrylate (DEA), oligo(ethylene glycol) methacrylate (OEGMA), 2-hydroxyethyl methacrylate (HEMA), glycerol monomethacrylate (GMA), 2-methacryloyloxyethyl phosphorylcholine (MPC), and a carboxybetaine-based methacrylate [CBMA]. Their chemical structures and literature references (which contain appropriate experimental details) are summarised in Table 1.

Table 1 *Chemical structures and relevant literature references of hydrophilic monomers that are amenable to aqueous/methanolic ATRP*

Monomer conversions were determined by ^{1}H NMR spectroscopy in all cases. Molecular weight distributions were assessed using THF GPC or aqueous GPC, respectively.

In most cases silica or alumina chromatography was used to remove the spent Cu catalyst after the ATRP syntheses.

3 Results and Discussion

Exploiting ATRP as an 'enabling' technology, we have recently synthesised a wide range of new, controlled-structure copolymers. These include: (1) branched analogues of 'Pluronic' non-ionic surfactants; (2) schizophrenic polymeric surfactants which can form two types of micelles in aqueous solution; (3) novel sulfate-based copolymers for use as crystal habit modifiers; (4) zwitterionic diblock copolymers, which may prove to be interesting pigment dispersants. Each of these systems is discussed in turn below.

3.1 Branched Analogues of 'Pluronic' Non-ionic Surfactants

Poly(alkylene oxide)-based (PEO-PPO-PEO) triblock and diblock copolymers are commercially successful, linear non-ionic surfactants which are manufactured by BASF and ICI. Over the last four decades, these block copolymers have been used as stabilisers, emulsifiers and dispersants in a wide range of applications. With the development of ATRP, it is now possible to synthesise semi-branched analogues of these polymeric surfactants. In this approach, the hydrophobic PPO block remains linear and the terminal hydroxyl group(s) are esterified using an excess of 2-bromoisobutyryl bromide to produce either a monofunctional or a bifunctional macro-initiator. These macro-initiators are then used to polymerise OEGMA, which acts as the branched analogue of the PEO block (see Figures 2 and 3).

A range of OEGMA-PPO diblocks and OEGMA-PPO-OEGMA triblocks have been synthesised by ATRP in alcohol/water mixtures with variable OEGMA content and relatively low polydispersities.[18] We are now beginning to investigate their aqueous solution properties. Initial results suggest that these semi-branched surfactants are reasonably surface-active but form micelles less

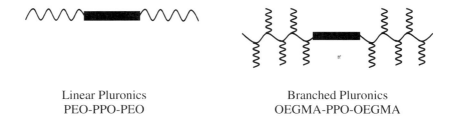

Linear Pluronics   Branched Pluronics
PEO-PPO-PEO   OEGMA-PPO-OEGMA

Figure 2 *Schematic representation of the linear and semi-branched architectures of two non-ionic polymeric surfactants*

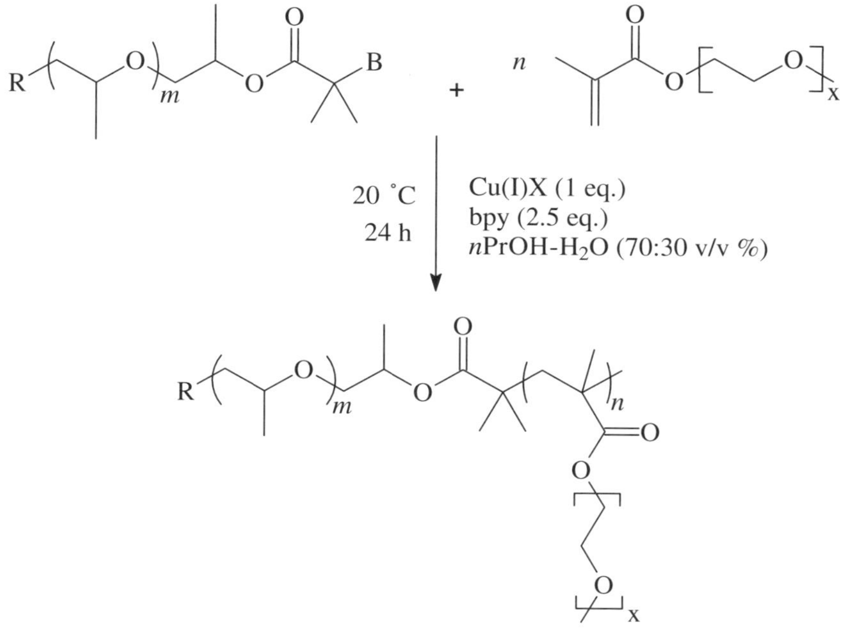

Figure 3 *Reaction scheme for the synthesis of a semi-branched OEGMA-PPO diblock copolymer by ATRP using a PPO-based macro-initiator*

readily than their linear counter-parts. Presumably, this is due to the poor packing efficiency of the branched OEGMA chains within the micelle corona. It remains to be seen whether these semi-branched surfactants offer any useful advantages over the conventional linear surfactants. We are also evaluating the possibility of preparing fully-branched analogues. In principle, the linear PPO block can be replaced with oligo(propylene oxide) monomethacrylate (OPOMA), but in practice we and others[19] have experienced difficulties in polymerising OPOMA with sufficient control by ATRP. To date, we have achieved high conversions but poor control over the molecular weight distribution; in most cases GPC analysis indicates multimodal traces.

3.2 Schizophrenic Polymeric Surfactants

It is well known that AB diblock copolymers form micelles in solvents that are selective for one of the blocks. By varying the nature of the solvent, it is also possible to form micelles with the A block in the core or with the B block in the core. However, we have recently demonstrated that certain hydrophilic AB diblock copolymers can form either A-core micelles or B-core micelles in aqueous media.[20-22] In the original example, both blocks were based on tertiary amine methacrylates and the diblock copolymer was prepared by group transfer polymerisation, a special type of anionic polymerisation which is particularly well-suited for methacrylic monomers. More recently, we have prepared a poly[propylene oxide-block-2-(diethylamino)ethyl methacrylate] (PPO-DEA) diblock copolymer by alcoholic ATRP using a macro-initiator strategy (see Figure 4).

The PPO block is temperature-sensitive, with an LCST (cloud point) of around 15 °C, whereas the DEA block is pH-responsive. At pH 6.5 and 5 °C, both blocks are solvated and the copolymer is molecularly dissolved in aqueous solution. On adjusting the solution pH to pH 8.5, the DEA block becomes deprotonated and hence hydrophobic, leading to the formation of DEA-core micelles at 5 °C. On the other hand, if the pH is held constant at pH 6.5 and the temperature is raised above the LCST of the PPO block, PPO-core micelles are obtained with the weakly cationic DEA chains forming the micelle coronas. Since these new diblock copolymers can exist in two micellar states we have christened them 'schizophrenic' copolymers.[12]

3.3 Synthesis of Sulfate-based Block Copolymers as Crystal Habit Modifiers for BaSO$_4$

We have recently synthesised a range of PEG-SEM diblock copolymers. PEG macro-initiators were used for the ATRP of SEM and the molecular weights of the PEG and SEM blocks were varied independently. In collaboration with Meldrum's group at QMC, these copolymers were evaluated as crystal habit modifiers for BaSO$_4$, which was prepared *in situ* by the addition of a soluble barium salt to an aqueous solution of (NH$_4$)$_2$SO$_4$. In the absence of PEG-SEM copolymer, the precipitated BaSO$_4$ particles had a polydisperse, micrometre-sized platelet morphology (see Figure 5a). In the presence of a PEG$_{45}$-SEM$_{23}$ diblock copolymer, egg-shaped particles were obtained (see Figure 5b). Presumably the sulfate groups in the SEM blocks interact with the crystal lattice of the

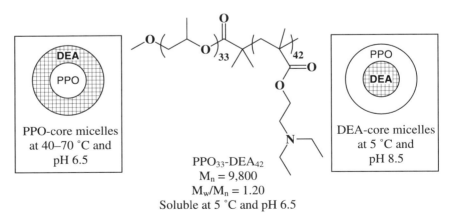

PPO-core micelles
at 40–70 °C and
pH 6.5

PPO$_{33}$-DEA$_{42}$
M_n = 9,800
M_w/M_n = 1.20
Soluble at 5 °C and pH 6.5

DEA-core micelles
at 5 °C and
pH 8.5

Figure 4 *Summary of aqueous solution conditions for the two micellar forms of the 'schizophrenic' PPO$_{33}$-DEA$_{42}$ diblock copolymer*

(a) (b)

Figure 5 *Scanning electron micrographs of (a) BaSO₄ particles prepared in the absence of any copolymer (control experiment) and (b) BaSO₄ particles prepared in the presence of a PEO₄₅-SEM₂₃ diblock copolymer synthesised by ATRP. Note the profound change in particle morphology due to the interaction of the sulfate-based block with the BaSO₄ crystal lattice*

precipitating $BaSO_4$ and the PEG blocks act as a steric stabiliser, leading to a profound change in the particle morphology. Control experiments with PEG and SEM homopolymers confirm that the diblock copolymer architecture is essential for effective crystal habit modification. Further characterisation of these egg-shaped particles using thermogravimetry, FT-IR spectroscopy and X-ray diffraction is in progress.[23] Similar crystal habit modification effects have been observed by Antonietti's group[24,25] for a range of carboxylate- and phosphonate-based copolymers; the advantage in the present study is the relative ease with which well-defined PEG-SEM diblock copolymers can be synthesised.

3.4 Synthesis of Zwitterionic Diblock Copolymers

There has been relatively little work on zwitterionic diblock copolymers, since the synthesis of these materials normally require protecting group chemistry and anionic polymerisation techniques.[26,27] However, it has been confirmed that these materials exhibit rich and complex phase behaviour in aqueous solution, including reversible precipitation at a certain critical pH, where the anionic charge just balances the cationic charge, *i.e.* the isoelectric point.[26–28] Moreover, in collaboration with a group at Akzo Nobel, Creutz and Jerome have shown[29] that a zwitterionic diblock copolymer based on methacrylic acid [MAA] and 2-(dimethylamino)ethyl methacrylate [DMA] appears to act as a ubiquitous dispersant for a range of inorganic and organic pigments with both anionic and cationic surface charge.

In principle, aqueous ATRP offers the tantalising possibility of the direct synthesis of reasonably well-defined zwitterionic block copolymers in water without recourse to protecting group chemistry. However, ATRP in acidic media is generally unprofitable, hence the (co)polymerisation of acidic monomers such as methacrylic acid or 4-vinylbenzoic acid must be carried out in weakly alkaline solution, *i.e.* the monomer should be in its anionic carboxylate

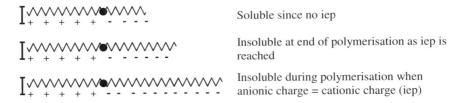

Soluble since no iep

Insoluble at end of polymerisation as iep is reached

Insoluble during polymerisation when anionic charge = cationic charge (iep)

Figure 6 *Schematic representation of the problems that can be encountered during the synthesis of a zwitterionic diblock copolymer by aqueous ATRP.* I *represents the initiator fragment and* • *represents the block junction*

form. Thus if quaternised monomers are used to synthesise zwitterionic block copolymers and the shorter block is prepared first, there is an obvious danger of passing through the isoelectric point during synthesis, leading to premature precipitation (see Figure 6).

We have some evidence that this theoretical problem is a genuine limitation in the case of a quaternised styrenic monomer which is block copolymerised with NaVBA.[30] This problem can be circumvented in two ways. Firstly, the polymerisation sequence can be simply reversed so that the longer block is synthesised first. If this is the quaternised block, the resulting copolymer cannot exhibit an isoelectric point because the major block is permanently cationic, thus no charge compensation can occur. On the other hand, if the longer block is anionic, then addition of HCl will protonate the acidic monomer residues and at some point an isoelectric point will be attained (unless the acidic block is strongly acidic, *e.g.* 4-styrenesulfonic acid).

Secondly, the quaternised monomer may be replaced with a weakly basic monomer such as MEMA, which exists in its neutral, non-protonated form in alkaline media. Thus the desired zwitterionic block copolymer is prepared in its anionic/neutral form so that no isoelectric point is encountered during the copolymer synthesis. Afterwards, the solution pH can be adjusted to the isoelectric point by the addition of acid to protonate the weakly basic MEMA residues and precipitate the copolymer, which might be a useful alternative approach to column chromatography for the efficient removal of the ATRP catalyst.

4 Conclusions

ATRP is a powerful synthetic tool for the synthesis of low molecular weight (Dp < 100–200), controlled-structure hydrophilic block copolymers. Compared to other living radical polymerisation chemistries such as RAFT, ATRP offers two advantages: (1) facile synthesis of a range of well-defined macro-initiators for the preparation of novel diblock copolymers; (2) much more rapid polymerisations under mild conditions in the presence of water. In many cases these new copolymers have tuneable surface activity (*i.e.* they are stimuli-responsive) and exhibit reversible micellisation behaviour. Unique materials such as new 'schizophrenic' diblock copolymers or sulfate-based copolymers for crystal habit modification are readily prepared by ATRP. Finally, zwitterionic diblock copolymers can be synthesised directly by aqueous ATRP without recourse to protecting group chemistry and may have interesting properties as ubiquitous pigment dispersants.

Two major challenges remain: (1) the more efficient and cost-effective removal (and preferably recycling) of the ATRP catalyst and (2) the extension of aqueous/methanolic ATRP to include other classes of monomers such as acrylates and (meth)acrylamides.

Acknowledgements

The following sponsors are thanked for their continued support of the water-soluble polymers research programme at Sussex: EPSRC, BBSRC, ICI Paints, Unilever, TotalFinaElf, GlaxoSmithKline, Syngenta, Avecia, Biocompatibles and Laporte Performance Chemicals.

References

1. J.S. Wang and K. Matyjaszewski, *J. Am. Chem. Soc.*, 1995, **117**, 5614.
2. M. Kato, M. Kamigaito, M. Sawamoto and T. Higashimura, *Macromolecules*, 1995, **28**, 1721.
3. T.E. Patten, J. Xia, T. Abernathy and K. Matyjaszewski, *Science*, 1996, **272**, 866.
4. J.H. Xia, S.G. Gaynor and K. Matyjaszewski, *Macromolecules*, 1998, **31**, 5958.
5. K. Matyjaszewski, Y. Nakagawa and C.B. Jasieczek, *Macromolecules*, 1998, **31**, 1535.
6. X.S. Wang, S.F. Lascelles, R.A. Jackson and S.P. Armes, *Chem. Commun.*, 1999, 1817.
7. K.L. Robinson, D.Phil; University of Sussex, UK, 2001.
8. X.S. Wang, R.A. Jackson and S.P. Armes, *Macromolecules*, 2000, **33**, 255.
9. E.J. Ashford, V. Naldi, R. O'Dell, N.C. Billingham and S.P. Armes, *Chem. Commun.*, 1999, 1285.
10. F.Q. Zeng, Y.Q. Shen, S.P. Zhu and R. Pelton, *J. Polym. Sci., Polym. Chem.*, 2000, **38**, 3821.
11. F.L.G. Malet, N.C. Billingham and S.P. Armes, *ACS Polym. Prepr.*, 2000, **41**(2), 1811.
12. S.Y. Liu, N.C. Billingham and S.P. Armes, *Angew. Chem., Int. Ed.*, 2001, **40**, 2328.
13. X.S. Wang and S.P. Armes, *Macromolecules*, 2000, **33**, 6640.
14. K.L. Robinson, M.A. Khan, M.V. de Paz Báñez, X.S. Wang and S.P. Armes, *Macromolecules*, 2001, **34**, 3155.
15. M. Save, J.V.M. Weaver and S.P. Armes, *Macromolecules*, 2002, **35**, 1152.
16. E.J. Lobb, I. Ma, N.C. Billingham, S.P. Armes and A.L. Lewis, *J. Am. Chem. Soc.*, 2001, **123**(32), 7913.
17. I. Ma, E.J. Lobb, S.P. Armes and N.C. Billingham, manuscript in preparation, 2001.
18. K.L. Robinson, M.V. de Paz Báñez, X.S. Wang and S.P. Armes, *Macromolecules*, 2001, **34**(17), 5799.
19. B.L. Sadicoff, A.E. Acar and L.J. Mathias, *ACS Polym. Prepr.*, 2000, **41**(2), 128.
20. V. Bütün, N.C. Billingham and S.P. Armes, *J. Am. Chem. Soc.*, 1998, **120**, 12135.
21. V. Bütün, N.C. Billingham and S.P. Armes, *J. Am. Chem. Soc.*, 1998, **120**, 11818.
22. V. Bütün, N.C. Billingham and S.P. Armes, *Macromolecules*, 2001, **34**, 1148.
23. K.L. Robinson, J.V.M. Weaver, S.P. Armes, E. Diaz Marti and F.C. Meldrum, *J.*

Mater. Chem., 2002, **12**(4), 890.

24. L. Qi, H. Cölfen and M. Antonietti, *Chem. Mater.*, 2000, **12**, 2392.
25. L. Qi, H. Cölfen and M. Antonietti, *Angew. Chem., Int. Ed.*, 2000, **39**, 604.
26. A.B. Lowe, N.C. Billingham and S.P. Armes, *Macromolecules*, 1998, **31**, 5991.
27. S. Creutz and R. Jerome, *Langmuir*, 1999, **15**, 7145.
28. S. Creutz and R. Jerome, *Prog. Org. Coatings*, 2000, **40**, 21.
29. S. Creutz, R. Jerome, G.M.P. Kaptijn, A.W. van der Werf and J.M. Akkerman, *J. Coatings Tech.*, 1998, **70**, 41.
30. S.A. Furlong, D.Phil., University of Sussex, UK, 2001.

Role of Colloidal Interactions in Determining Rheology

D.M. Heyes and J.F.M. Lodge

DEPARTMENT OF CHEMISTRY, UNIVERSITY OF SURREY, GUILDFORD, GU2 7XH, UK

1 Introduction

Colloidal liquids are widespread in the natural world and in manufactured processes and products (*e.g.* foods, paints and other coatings, agrochemicals, ceramics and polymers). The macroscopic behaviour of these liquids is largely governed by the colloidal particle interactions and it is therefore on this scale that a true understanding of these materials must be sought. The relationships between the physical properties of colloidal liquids and the interactions between the particles are still not well understood, especially when there are strong attractive interactions between the particles. The complexity of these systems, together with the added complication of hydrodynamic interactions between the particles, means that analytic treatments of their dynamical and physical properties are highly approximate. There has been only limited progress in quantifying the effects of chemical sensitivity on the physical properties. Situations of practical importance also often employ high concentrations, sometimes up to close-packing, in which particle 'crowding' factors are also important. Molecular simulation is a useful quasi-experimental tool for exploring these issues. Molecular simulation and perturbation theory have been used in this study to achieve an improved understanding of the relationship between the particle interactions and the molecular relaxation processes occurring in these liquids, and in particular their impact on the linear viscoelasticity. We have considered some basic interaction types which are found in colloid systems.

2 Exploring the Potentials

The systems considered included a steeply repulsive potential, SRP, between the particles with n arbitrary:

$$V(r) = \varepsilon \left(\frac{\sigma}{r}\right)^n \tag{1}$$

These are purely repulsive particles. An attractive potential of the Lennard–Jones type, with n arbitrary:

$$V(r) = 4\varepsilon\left[\left(\frac{\sigma}{r}\right)^{2n} - \left(\frac{\sigma}{r}\right)^{n}\right] \qquad (2)$$

was also considered. These particles have an attraction at intermediate and long range, which enables the particles to aggregate and even form network or gel-like states at low concentrations.

2.1 Near-hard Sphere Fluids – a Step Towards Stabilised Colloids

The steeply repulsive particles, SRP, whose interaction potential is given in Equation (1), offer an extremely useful starting point for understanding the rheology of colloidal liquids because of the versatility of this potential in representing a wide range of stabilised systems. Equation (1) has the useful feature that in the large n limit ($n \to \infty$) we approach the hard sphere potential. Colloid scientists talk of 'hard sphere' colloids, and this potential can explore both this limit and much softer repulsions (down to $n = 1$, the Coulomb interaction, in principle) just by adjusting a single parameter (n). Using molecular dynamics computer simulation and statistical mechanics, some fundamental characteristics of the viscoelasticity of these fluids have been uncovered.[1,2] The shear stress relaxation function, $C(t)$ (the response of the shear stress at time t after a step in shear strain at time $t = 0$) is the key quantity in defining viscoelastic behaviour. It was shown that for the SRP fluids (without the solvent) there is an accurate expansion of $C(t)$ about $t = 0$ for n large. For $C(t)$ (normalised so that $C(0) = 1$) we find,

$$C(t) = 1 - x^2 + \ldots O(x^4) \qquad (3)$$

where $x = n(k_B T/\varepsilon)^{1/2}(\varepsilon/m\sigma^2)^{1/2}t$ and m is the mass of the SRP particle. The parameter, x, is a non-dimensional time that includes the temperature and the potential stiffness, n. Note that the stress relaxation function is independent of density, which is not surprising as in the large n limit, the short time decay of $C(t)$ is dominated by binary collisions at all densities, even near close packing. (Three and higher body effects are required to introduce a density sensitivity.) It is tempting to extrapolate this series by performing the following closure, which might be applicable at all times:

$$C(t) = \frac{1}{1 + x^2} = 1 - x^2 + x^4 - O(x^6) \qquad (4)$$

The function fits the $C(t)$ data very well.[1,2] Note that the n-dependence of the reduced time parameter x means that the relaxation time, τ (which is the area under $C(t)$) decreases as n^{-1}. The dependence of the viscosity, η, with n is much weaker however as $\eta = G_\infty \tau$, where G_∞ is the infinite frequency elastic modulus, which is proportional to n for large n. The two n dependencies do not exactly cancel out in practice, as $C(t)$ has contributions from cage effects at long time which cause deviations in the form of $C(t)$ from the analytic expression given in Equation (4). Nevertheless, the n-dependence of the viscosity is much less that the G_∞ and τ separately. In Figure 1 we show an example of the short-time scaling of $C(t)$ when plotted against nt, with $n = 144$ and 1152 as examples, and when compared also with Equation (4). The effective hard-sphere packing fraction is 0.45 and the reduced temperature is $kT/\varepsilon = 1$ in this figure.

One conclusion from this study is that although the hard-sphere fluid has been very successful as a reference fluid, for example, in developing analytical equations of state, it is unrealistic in representing the dynamical relaxation processes in real systems, even with very steeply repulsive potentials. Owing to the discontinuity in the hard-sphere potential, this fluid, in fact, is not a good reference fluid for the short time (fast or 'β') viscoelastic relaxation aspects of rheology.

2.2 Gel Formation and Ageing – the Role of Particle Attraction Strength and Range

Despite its importance in many consumer products, the physical origins of the gel state, and its dependence on the chemical forces between the particles, is still quite poorly understood. Various approximate theoretical models exist. For example, the fractal-based approach has had considerable success in fitting to a wide range of experimental data, but only over certain parameter ranges – for

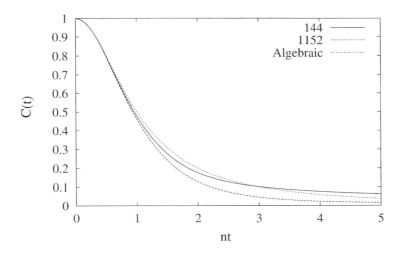

Figure 1 *The shear stress relaxation function, C(t), obtained from a molecular dynamics simulation of 500 SRP spheres at a reduced temperature of 1.0 and effective volume fraction of 0.45. Note that n = 144 and 1152 (from Equation (1)) cases are superimposable with the analytic function of Equation (4) ('Algebraic' on the figure) for short times, t (or nt here)*

strong particle attractions and at particle densities close to the percolation threshold. The fractal model for rheology makes rather sweeping assumptions about particle and cluster dynamics, and is therefore only applicable in the time and space 'scale-invariant' regimes close in time to the percolation threshold itself. The early success of this approach has in many ways stifled further theoretical work in this area, even though there is no specific chemical focus in these treatments. One can view the gel state as a 'quench' into the two phase region of the phase diagram ('vapour–solid' or 'vapour–liquid' states of the associating particles) which leads to the formation of a percolating network, substantial condensation of the assembly, and local structure evolution can take place, even for a system with a fractal dimension of *ca.* 3. The 'constituency' of the gel depends on a range of factors, such as the strength and range of the attraction between the particles, as well as their concentration.

Brownian dynamics, BD, simulations were carried out on model colloidal liquids interacting with the potential given in Equation (2).[3] The structural evolution and rheology of these model attractive spherical colloidal particles were followed as they self-assembled into long-range networks. The procedure used was to 'quench' the particles from a supercritical state point into the vapour–liquid or vapour–solid parts of their phase diagrams. The solids volume or packing fractions, ϕ, were in the range 0.05–0.20. The interactions between the model colloidal particles had $n = 6$, 12 and 18 in Equation (2), giving rise to 12:6, 24:12 and 36:18 potentials, respectively. Along this series, the attractive part of the potential becomes shorter ranged. These systems developed a gel-like morphology during the simulations, with the aggregate morphology and rheology being sensitive to the range of the attractive part of the potential and the position in the phase diagram of the quench. A typical example is shown in Figure 2, for the long-range 12:6 potential, which self-assembled on intermediate timescales to compact structures with thick filaments. The systems generated using the shorter-ranged 24:12 and 36:18 potentials persisted in a more diffuse network for the duration of the simulations and evolved more slowly with time.

The phase separation also becomes apparent in a time-evolving pair radial distribution function, $g(r)$, an example of which is shown in Figure 3 for a quenched 24:12 system. With time after the quench, the $g(r)$ shows the appearance of a second and even third coordination shell. Notice that the second coordination shell peak is actually formed from two overlapping peaks, which is a signature typical of the glassy state. The condensation of the particles with time is also manifest in the average interaction energy per particle, u, which is shown in Figure 4 for various packing fractions (0.05 to 0.20) as given on the figure. The denser systems form more compact structures and therefore develop the most negative energy values for u with time.

The rheology of many of the systems displayed gel-like viscoelastic features, especially for the *long-range* attractive interaction potentials, which manifested a non-zero plateau in the shear stress relaxation function, $C_s(t)$, the so-called equilibrium modulus, G_{eq}, which has been considered to be a useful indicator of the presence of a gel. The infinite frequency shear rigidity modulus, G_∞ was extremely sensitive to the form of the potential. Despite being the most short-

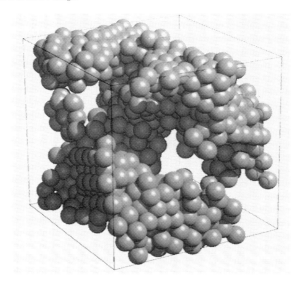

Figure 2 *Snapshot picture depicting the aggregate structure at* $ta^{-2}D_0 = 164$ *(where* a *is the particle radius and* D_0 *is the self-diffusion coefficient of the particles at infinite dilution). There are 864 particles in the simulation with particles that interact with the 12:6 potential at the state point,* $\phi = 0.2$ *and* $T^* = 0.3$. *Notice the filamentary nature of the agglomeration of the particles*

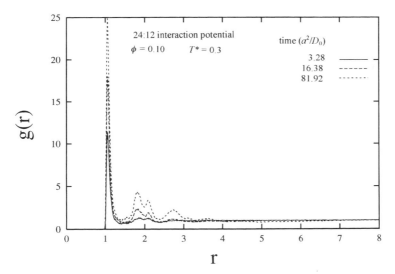

Figure 3 *The* g(r) *at various* $ta^{-2}D_0$ *(shown on the figure) for the 24:12 interaction potential, with* $\phi = 0.1$ *and* $T^* = 0.3$

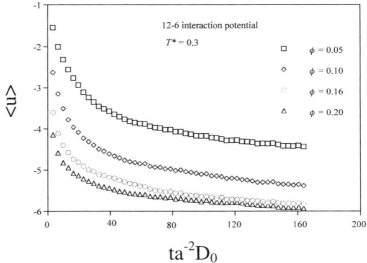

Figure 4 *Time evolution of the average interaction energy,* u, *for the 12:6 interaction at*
T* = 0.3 *for different volume fractions, which are given on the figure*

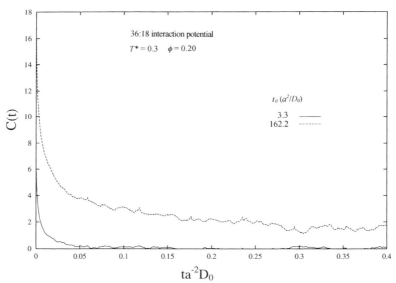

Figure 5 *Time evolution of the shear stress relaxation function,* C(t), *for the 36:18 potential*
at ϕ = 0.2 *and* T* = 0.3. *The waiting times are* t_w = 3 *and 162 for the two curves*

lived, the 12:6 potential systems gave the most pronounced gel-like rheological
features.

In common with glasses, the dynamics of particle gels can be strongly dependent on its history of formation. One finds that the relaxation dynamics become
increasingly slower with 'age' or 'waiting' time from the quench, t_w. The stress
relaxation function now depends on *two* times, $C(t_w, t + t_w)$. The larger t_w, the

slower the function decays up to times of order $2t_w$. In fact, for Lennard–Jones glasses the long time behaviour of the function can be scaled quite well onto a single curve using the reduced time parameter t/t_w.[4,5] One would expect the nature of the aging process in transient gels to be different from the glassy case because of the accompanying long-range morphological changes, which are additional important factors in the ageing process. Parallel ageing takes place on a range of lengthscales, much larger than in the case of glasses. We found this ageing effect occurred in our systems. It was manifest as a 'plateau' in $C(t)$ at intermediate times, which grew in height and duration with t_w. An example of this behaviour is given in Figure 5 for the 36:18 potential system at a packing fraction of 0.20 and reduced temperature of 0.3. Note that after a time of 162 reduced time units after the quench, the shear stress relaxation function has developed an intermediate plateau, characteristic of ageing systems.

3 Conclusions

The research carried out on these very simple potential forms has revealed that they capture many of the physical/rheological features present in real systems. By focusing on such potentials, which have one or two adjustable parameters one can gain insights into generic effects that span many examples of chemical system. These potentials are also, on occasions (*e.g.* the potential given in Equation (1)) amenable to simple analytic treatments.

Acknowledgements

The authors would like to thank the Engineering and Physical Sciences Research Council of Great Britain (EPSRC) for funding this work.

References

1. J.G. Powles and D.M. Heyes, *Mol. Phys.*, 2000, **98**, 917.
2. J.G. Powles, G. Rickayzen and D.M. Heyes, *Proc. Roy. Soc.: Series A*, 1999, **455**, 3725.
3. J.F.M. Lodge and D.M. Heyes, *Phys. Chem. Chem. Phys.*, 1999, **1**, 2119.
4. J.-L. Barrat and W. Kob, *Europhys. Lett.*, 1999, **46**, 637.
5. J.-L. Barrat and W. Kob, *J. Phys.: Condens. Matter*, 1999, **11**, A247.

New Polymeric Dispersants for Very Fine Calcium Carbonate Slurries

Y. Duccini and A. DuFour

ROHM AND HAAS FRANCE SAS, EUROPEAN LABORATORIES, SOPHIA ANTIPOLIS, 06560 VALBONNE, FRANCE

1 Introduction

Calcium carbonate is the most widely used filler mineral in the world. Its success is mainly due to its availability and also its suitability for a large range of applications.

Many industries use calcium carbonate:

* Paper industry (as filler and for coating)
* Paint industry (as filler and extender)
* Plastic industry (as filler, but bringing interesting properties)
* Sealant and adhesive industry (as filler)

Each application requires very specific product characteristics in terms of:

* Chemical purity
* Particle size distribution
* Particle shape and surface area
* Whiteness
* Rheological behaviour

There are two sources of calcium carbonate, ground calcium carbonate (GCC) and precipitated calcium carbonate (PCC).

Ground calcium carbonate, extracted from the earth, is present in practically every country in the world in varying quantities in the form of limestone, marble, dolomite or chalk. Following the extraction, GCC needs to be ground. Dry grinding, the cheaper alternative, is often limited to a minimum particle size of 2–3 microns. Wet grinding, more expensive, is used for fine and ultra fine material or when the final product must be a slurry (paper or paint application).

Precipitated calcium carbonate is produced by chemical reaction between quick lime and carbon dioxide. By changing the reaction conditions, numerous types of PCCs can be produced, differing in terms of particle size, particle shape, crystal type, *etc*. PCC is generally not ground. It can be sold as dry powder or as slurry.

When calcium carbonate (GCC or PCC) is sold in a liquid form, the concentration of solids in the slurry is maximised to minimise transportation of water and to reduce energy consumption for drying in some applications.

The way to maximise the solids content of the slurry while keeping a reasonable viscosity is through the use of a dispersant. The dispersant helps maintain fine solid particles in a state of suspension, thus minimising their agglomeration or settling. The dispersant is generally a low molecular weight acrylic based polymer. For most applications, cost of dispersant is critical.

The paper industry requires the finest grade of calcium carbonate for coating of the paper sheet. The brightness of the paper sheet improves with increasing fineness of calcium carbonate in the coating colour. In addition, coating machines operating at very high speed demand a very good rheological behaviour of the coating colour, primarily low viscosity at high speed.

These specific requirements of a calcium carbonate slurry used for paper coating have necessitated the development of new dispersants that allow the preparation of ultra fine calcium carbonate slurries at high solids concentration that provide good rheological behaviour in coating.

This article studies some of the polymer characteristics influencing the slurry dispersion and describes a new class of acrylic based polymer showing promising dispersing properties for ultra fine calcium carbonate slurries.

2 Acrylic Based Dispersants Manufacturing

These polymers are synthesised in water by free radical polymerisation. In the reaction, many components are involved, principally:

- The monomers (M, M′, *etc*.) that will create the backbone of the polymer.
- The initiator (R–R) that will provide free radicals to start the polymerisation reaction.
- The chain transfer agent (X–A) that will regulate the molecular weight of the polymer.

The process is schematically the following when using only one monomer (acrylic acid):

Initiation:

$$R-R \rightarrow 2\,R^{\cdot}$$
$$R^{\cdot} + M \rightarrow RM^{\cdot}$$

Propagation:

$$RM^{\cdot} + M \rightarrow RM_2^{\cdot}$$

Regulation:

$$RM_n{}^{\cdot} + X{-}A \rightarrow RM_n{-}X + A^{\cdot}$$
$$A^{\cdot} + M \rightarrow AM^{\cdot}$$
$$AM^{\cdot} + M \rightarrow AM_2{}^{\cdot}$$

Termination:

$$RM_p{}^{\cdot} + RM_q{}^{\cdot} \rightarrow RM_{(p+q)}R$$

In the industry, the polymerisation of more than one monomer gives a random copolymer, monomers being distributed along the chain without any order.

Many different chain transfer agents have been used to produce such polymers. These are mercaptans (*e.g.* mercaptopropionic acid), alcohols (*e.g.* isopropanol), or inorganic compounds (*e.g.* bisulfite, hypophosphite, *etc.*).

The chain transfer agent is critical for production of low molecular weight polymers and can also influence performance properties.

3 Acrylic Based Dispersant Characterisation

Acrylic based polymers can be characterised by:

- Molecular weight (M_w and M_n)
- Polydispersity (defined as M_w/M_n)
- Monomer type
- Monomer ratio
- End group
- Neutralising agent

To that, we have to add the concentration of salts of the resulting aqueous solution of polymer. These salts come from the residual polymerisation auxiliaries and the neutralising agents.

4 Influence of Polymers Characteristics on Slurry Dispersion

A good dispersant must allow the preparation of slurries that provide the highest solids content with the lowest (or at least acceptable) viscosity.

For paper coating the paper industry increasingly requires very fine particle size and high solids calcium carbonate slurries. Today, it is normal to prepare slurries at 78% solids with 90% (and even more) particles less than 2 microns. Fine particle size calcium carbonate provides high brightness of the paper sheet, while a high solids slurry allows improved paper coating quality (less mottling, better gloss, better drying, *etc.*). These very fine particle size slurries require the use of improved dispersants. The previously mentioned polymer characteristics play an important role on dispersion efficacy. Depending on the type of mineral and its particle size, there will be an optimum molecular weight allowing a good dispersion.

We know that commercial polymers do not have a single molecular weight but have a range around an average value. This range is evaluated by the polydispersity. The lower the polydispersity, the narrower the range. Low polydispersity is beneficial since there are fewer low molecular weight polymeric chains that are inefficient dispersants and fewer high molecular weight polymeric chains that can act as flocculants.

The monomer type and ratio change the adsorption properties of the polymer on the calcium carbonate particles. The type of end group (which depends on the polymerisation process) can also influence adsorption characteristics.

The neutralising agent as well as the salt content influences the slurry ageing. A good dispersing agent must maintain constant viscosity over time. We know that a lower monovalent cation content improves slurry stability.

The influence of molecular weight, polymerisation process, monomer type and neutralisation have been studied.

5 Calcium Carbonate Tested

Three different sorts of calcium carbonate have been used for tests, one GCC and two PCCs.

- The first PCC has a particle size of 70% < 2 microns. The slurry was prepared at 75% solids, and the dispersant concentration was 0.80% (as polymer solids/dry $CaCO_3$).
- The second PCC is finer, its particle size is 95% < 2 microns. Slurry solids was also 75%, and dispersant concentration was 0.80% (as polymer solids/dry $CaCO_3$).
- The tested GCC was ground to a particle size of 90% < 2 microns. The slurry concentration was 74.5%, and dispersant concentration was 0.73% (as polymer solids/dry $CaCO_3$).

The following measurements have been performed on PCC slurries:

- pH
- Solids
- Brookfield viscosity at 10, 50 and 100 rpm
- Ageing test at $T = 0$, $T = 24$ hours and $T = 48$ hours

On GCC slurry, we measured:

- pH
- Solids
- Conductivity
- Particle size distribution
- Brookfield viscosity at 10, 20, 50 and 100 rpm
- Ageing test at $T = 0$ up to $T = 5$ days

Brookfield viscosity measurement is a way to estimate the quality of a fresh dispersion. The ageing test controls the evolution of the viscosity over time. It is important to maintain low viscosity for several days, in order to keep the slurry workable.

The ageing test is particularly important for fine particle size slurries since they tend to gel when an inappropriate dispersant has been used.

The ageing test measures the torque necessary to induce the rotation of a special spindle in the slurry. The measure must be done before the first rotation is completed in order to prevent destruction of the gel network. We measure gel after 25 seconds using a Brookfield viscometer with a special spindle (Helipath type), rotating at 1 rpm.

Particle size distribution was measured with Coulter LS 230 laser equipment. Solids were determined with a drying scale.

For each slurry, solids were corrected ($\pm 0.1\%$) prior to viscosity measurements. For GCC grinding, the tolerance on particle size distribution was $\pm 1\%$ (89–91% < 2 microns).

6 Influence of Polymerisation Process

Rohm and Haas developed a new polymerisation process based on the use of a *new chain* transfer agent (*process C*). Polymers made with this new process showed very good dispersing activity.

Polymers prepared by current industrial processes have been compared with this new process in the dispersion of GCC and PCCs (Tables 1–3). These current processes use as chain transfer agent *inorganic salts* (*processes A and B*) or alcohol (*process D*).

Table 1 *Fine PCC (95% < 2 microns), slurry solids 75%, dispersant 0.80% (as polymer solids/dry powder)*

Sample	Process	Viscosity Brookfield 100 rpm (mPas)	Ageing After 48 hour (/10^6 mPas)
1	A	1072	20.9
2	B	1158	15.7
3	C	614	10.3
4	D	582	15.2

Table 2 *Coarser PCC (70% < 2 microns), slurry solids 75%, dispersant 0.80% (as polymer solids/dry powder). Process A was not tested with this PCC*

Sample	Process	Viscosity Brookfield 100 rpm (mPas)	Ageing After 48 hour (/10^6 mPas)
1	B	604	9.1
2	C	444	5.5
3	D	482	9.2

Table 3 *GCC, slurry solids 74.5%, dispersant 0.73% (as polymer solids/dry powder), grinding at 90% < 2 microns*

Sample	Process	Viscosity Brookfield 100 rpm (mPas)	Ageing After 5 days (/10^6 mPas)
1	A	318	138
2	B	242	138
3	C	166	39
4	D	160	37

The average molecular weight (M_w) of the tested polymers is 4000 and polydispersity is about 1.2.

From the three series of tests, we can rank the polymer processes in term of dispersing efficacy for calcium carbonate as follows:

$$A < B < D \leq C$$

Since we tested industrial polymer samples which may have differences apart from the chain transfer agent, we cannot conclude that the difference in efficiency is only due to the chain transfer agent.

7 Influence of Dispersant Molecular Weight

The influence of the molecular weight of the dispersant on the quality of the slurry dispersion has been studied on both PCCs. The dispersants ranged in molecular weight from 2000 to 5800 (according to our internal standards) made by the *process C*. Results are summarised in Tables 4 and 5, below.

We clearly see that for both PCC qualities, the optimum molecular weight for the dispersant is around 5000.

Ageing is better for the coarser PCC grade. The observation that the finer the calcium carbonate particle size, the higher the tendency to gel is likely due to fine particles having greater surface area, thus having inadequate dispersant to cover the surface and provide particle stability.

The ageing test indicates that the higher the molecular weight, the worse the ageing behaviour. In both cases, the lower molecular weight dispersant gives the least viscosity increase. The long polymeric chains probably create a three-dimensional network by particle bridging which cause some slurry gelation.

There will be a compromise between the initial viscosity and the ageing behaviour in the choice of the best dispersant.

8 Influence of Monomers

The monomers constituting the backbone of the polymer influence the adsorption on particles and thus dispersion. Two polymers, prepared by the *process C*,

Table 4 *Fine PCC (95% < 2 microns), slurry solids 75%, dispersant 0.80% (as polymer solids/dry powder)*

Sample	Process	M_w	Viscosity Brookfield 100 rpm (mPas)	Ageing After 48 hours (/10^6 mPas)
1	C	2000	1086	8.5
2	C	3600	614	10.3
3	C	4880	512	17.9
4	C	5170	394	19.3
5	C	5800	430	23.4

Table 5 *Coarser PCC (70% < 2 microns), slurry solids 75%, dispersant 0.80% as polymer solids/dry powder)*

Sample	Process	M_w	Viscosity Brookfield 100 rpm (mPas)	Ageing After 48 hours (/10^6 mPas)
1	C	3600	444	5.5
2	C	5170	344	11.8

Table 6 *Effect of monomer composition on fine PCC dispersion*

Sample	Process	M_w	Viscosity Brookfield 100 rpm (mPas)	Ageing After 48 hours (/10^6 mPas)
1	AA	2000	1086	8.5
2	AA/X	2000	523	8.7

differing only by their monomer composition were tested on fine PCC dispersion (Table 6).

We see that the second monomer greatly improves the dispersing effect. Here again we can note that a low molecular weight polymer gives a good ageing behaviour.

9 Influence of Polymer Neutralisation

The influence of salt content in the polymer solution, and particularly monovalent salts used to neutralise the polymer, were also studied.

Fully neutralised and 60% neutralised polymers from the different processes were tested as dispersants for GCC grinding. Conditions were 74.5% solids slurry, 0.73% dispersant (as solids/dry powder) and 90% particle < 2 microns. Results are summarised in the Table 7.

By using partially neutralised polymers, we increase the adsorption of the polymer on the calcium carbonate particle surface. More polymer being adsor-

Table 7 *Effect of dispersant neutralisation on GCC dispersion*

Sample	Process	% Neutral	Viscosity Brookfield 100 rpm (mPas)	Ageing After 5 days (/10^6 mPas)
1	A	100	318	138
2	A	60	182	11.4
3	B	100	242	138
4	B	60	204	12.8
5	C	100	166	39
6	C	60	158	8
7	D	100	160	37
8	D	60	204	9

bed and less remaining in solution, dispersion is improved and gelation is reduced.

Dispersion is particularly improved by using partially neutralised polymers with standard dispersants like *processes A and B* since, due to their relatively poor efficiency, adsorption was far from complete with 100% neutralised dispersant amount. With *processes C and D*, dispersion improved very little although gelation tendency was reduced.

10 Special Purpose Dispersant

We developed another dispersant presenting properties for a special application. In some cases when an ultra fine calcium carbonate slurry is spray dried, a re-agglomeration of the powder often occurs, due to high water adsorption. This powder aggregation is detrimental to the quality of the final product.

Hygroscopicity tests were performed on slurries prepared with different dispersants after drying. Figure 1 shows that the water re-adsorption after drying is significantly reduced with the polymer called Acusol 460 N.

Acusol 460 N is an acrylic based copolymer containing a non-ionic monomer bringing a hydrophobic character which decreases the water re-adsorption of the spray dried powder.

11 Conclusion

With the paper industry requiring more and more sophisticated products from mineral producers and particularly finer and finer calcium carbonate particle size slurries, the need to improve dispersants is becoming a reality. The work described in this paper is the beginning of research on improved dispersants for preparation of PCC and GCC slurries and also other minerals.

The new route (*process C*) for manufacturing acrylic based polymers seems very promising. Polymers prepared by this way attain the efficiency of the best dispersants on the market. In the future, we will explore this route, improving parameters which are known to influence dispersant efficiency such as molecular

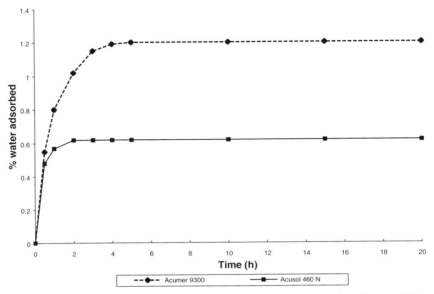

Figure 1 *Water re-adsorption of dried calcium carbonate prepared with two different dispersants*

weight, polydispersity, monomer type, monomer ratio, *etc.*

Understanding the adsorption mechanism of the polymer on calcium carbonate particles is also something important which should help the design of an improved dispersant. Increasing the level of adsorption of the polymer on particles will greatly enhance the dispersing efficiency.

Some special applications require specific dispersants. The case described in this paper shows that adding a property like hydrophobicity to a dispersant can help solve a problem in the final product. This is something we should not neglect when helping our customers to improve the quality of their products.

AMP-95™ for Mineral Dispersion

Thomas Gole

ANGUS CHEMIE GMBH, A SUBSIDIARY OF THE DOW CHEMICAL COMPANY, ZEPPELINSTRAßE 30, D-49479 IBBENBÜREN, GERMANY

1 Introduction

Dispersion is a term for systems containing various phases of at least one continuous and one finely dispersed. Referring to mineral slurries, this is typically a suspension of a mineral in water. This suspension normally contains some more additives for improved stability. One important additive in these systems is the dispersant. Interparticle forces hold the particles together and these interactions are reduced by the use of dispersants. This can be indicated by improved rheology profiles.

Dispersants function through various mechanisms. For water-based systems the preferred mechanism is stabilisation by ionic repulsion. A repulsion force layer is formed around the mineral particle. To maintain the suspension stability, the thickness of this layer around each particle has to be increased with increasing particle size. Layer decay is more frequent with the use of small particles, which results in higher proneness to partial flocculation. Also a uniform layer is necessary for effective stabilisation of all dispersed particles. AMP-95™ helps to achieve all these requirements.

AMP-95™ is a non-polymeric charge stabiliser; because of its organic structure and low molecular weight of 89 it is often referred to as a micro dispersant (Figure 1).

2 The Concept of Charge Stabilisation

Each pigment/mineral has a surface charge. This charge is dependant on various factors, among others the chemical structure of the mineral itself, the shape, geometry and particle size, but also the surface treatment. The surface charge is zero at the so-called isoelectric point (IEP) which relates to a very specific pH. The IEP of most pigments is around a pH of 3–7. A highly adsorbing alkaline agent, such and AMP (Figure 2), stabilises the charge and therefore the electro-

47

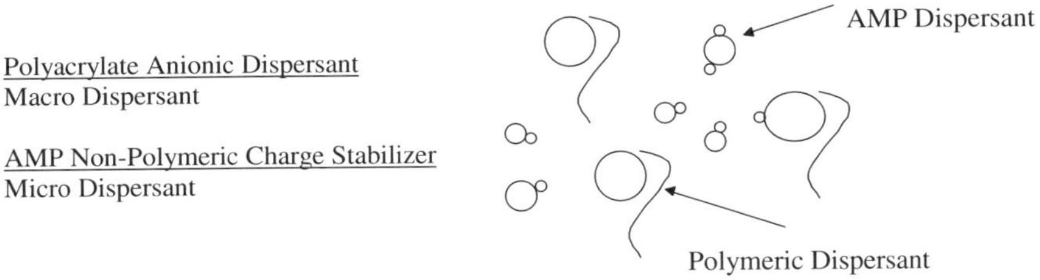

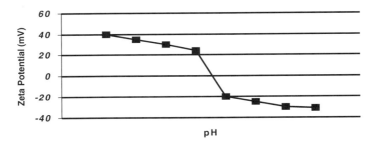

Polyacrylate Anionic Dispersant
Macro Dispersant

AMP Non-Polymeric Charge Stabilizer
Micro Dispersant

AMP Dispersant

Polymeric Dispersant

Figure 1 *Schematic representation of polymeric and AMP dispersant mode of action*

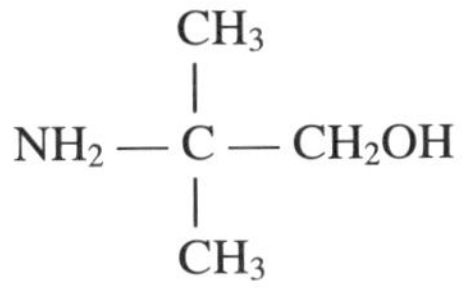

$$NH_2 - \underset{\underset{CH_3}{|}}{\overset{\overset{CH_3}{|}}{C}} - CH_2OH$$

Figure 2 *Molecular structure of AMP-95^{TM}*

Figure 3 *Plot of zeta potential vs. pH for a typical mineral suspension*

static repulsion. The pH is higher than the IEP and a negative charge and potential is developed (Figure 3).

3 AMP-95™ Properties

AMP's main use as a dispersant is due to the effect of ionic repulsion. Its functionality is to shift the pigment/filler surface charges to above the isoelectric point and thereby rendering the electrostatic repulsion of the pigment particles. When two particles come close together the diffuse layers penetrate each other and thereby generate the repulsion effect. The repulsion force has to be greater than the attracting van der Waals forces to maintain dispersibility. Typically the thickness of the diffuse layer has to be increased with increasing particle size because van der Waals forces also increase. In addition, the layer decays more easily when the particle size is small, and the system is more susceptible to flocculation. Micro dispersants such as AMP are very effective in this respective area.

Also, a uniform layer is necessary for effective stabilisation of all dispersed particles. AMP is small enough to penetrate easily to the pigments surface and adhere strongly by adsorption/neutralising acidic moieties. It is small enough not to form double coordination, as longer molecules are able to do. The branched structure (hydrophobic side chain) assists in this. Not only, as stated above, is the electron density enhanced; the side chains also are thought to prevent intermolecular hydrogen bonding. With many other dispersants, synergistic effects are noticed.

4 AMP-95™ Adsorption and Viscosity Profile

Adsorption isotherms for AMP-95™ onto kaolin, titania and calcium carbonate are shown in Figure 4. Viscosity profiles obtained when AMP is used to disperse titania and kaolin are shown in Figures 5–7.

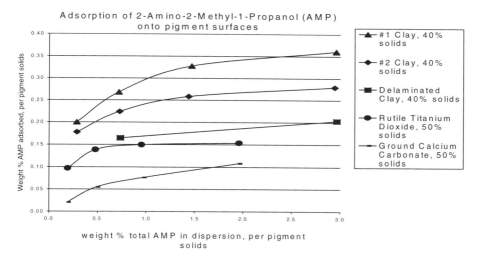

Figure 4 *Adsorption isotherms for AMP-95^{TM} onto kaolin, titania and calcium carbonate*

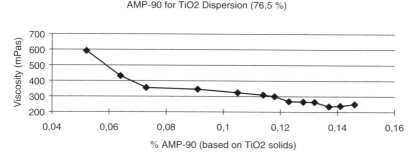

Figure 5 *Viscosity* versus *dose for titania dispersed with AMP*

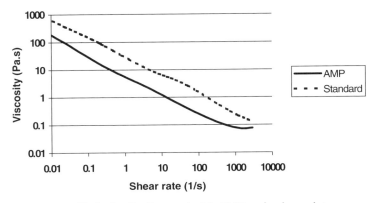

Figure 6 *Viscosity profile for kaolin dispersed with AMP and polyacrylate*

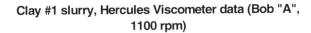

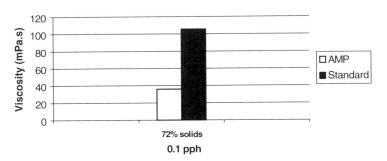

Figure 7 *Hercules viscosities for AMP and polyacrylate dispersed kaolin*

Table 1 *Colour details*

Formulation 1	pph
No. 1 Clay	70
CaCO$_3$	30
Dispersant	0.00–0.60
Rheology additive	Test
Latex	8
Starch	4
Thickener	Not used
% Solids	64.5–67%

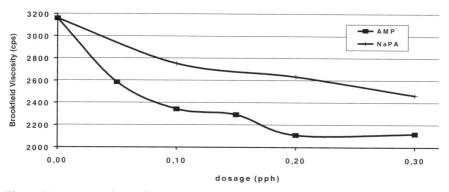

Figure 8 *Dispersant demand curve at low shear rate (Brookfield 30 rpm)*

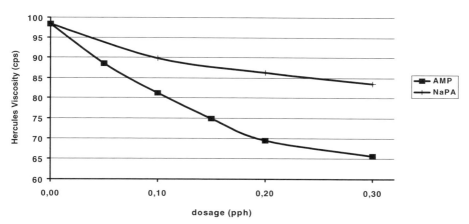

Figure 9 *Dispersant demand curve at high shear rate (Hercules, bob E, 4400 rpm)*

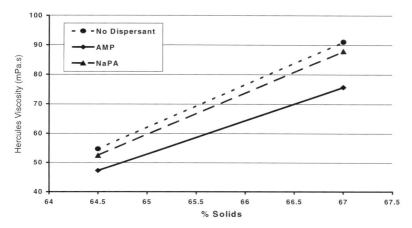

Figure 10 *Hercules viscosity versus solids plot for 70/30 Clay/CaCO₃ colour*

5 Example of the Use of AMP-95™ in a Typical Application – Paper Coating

The use of AMP as a rheology additive in paper coating is illustrated below. Table 1 gives formulation details. Figure 8 shows the colour viscosity at low shear rates and Figure 9 shows the colour viscoisty at high shear rates. The application was board basecoating at 62.5% solids.

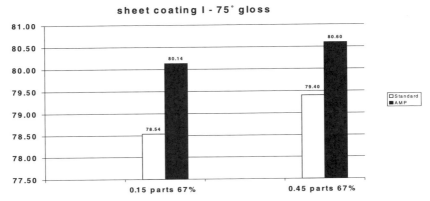

Figure 11 *Effect of AMP and standard dispersant on gloss for 70/30 Clay/CaCO$_3$ coating*

Table 2 *Effect of AMP on sheet gloss for 50/50 clay/carbonate coating*

Component	pph	pph
Clay	50	50
CaCO$_3$	50	50
Dispersant	0.15	0.15
AMP-95/Rheology Additive	0	0.2
Latex	13.5	13.5
Acrylic Thickener	0.3	0.3
Optical Brightner	1.5	1.5
Ca Stearate	0.8	0.8
Melamine Formaldehyde	1	1
% solids	66	66
Property	*Standard*	*With AMP-95*
PH	8.90	9.00
Brookfield @ 100 rpm	1612	1490
Hercules @ 4400 rpm	53.9	54
Brightness (%)	89.7	90.5
Gloss	55.6	58.6
Whiteness	87.6	88.1
L*	94.8	94.6
b	−0.17	−0.37
K&N (%)	9	9

In a typical paper coating formulation with 67% solids containing a 70/30 Clay/CaCO$_3$ combination AMP-95 showed improved rheology performance especially at higher shear rates as indicated by Hercules viscosity (Figure 10) and also improved the gloss at 75° (Figure 11).

Even in a 50/50 clay/CaCO$_3$ combination (with the CaCO$_3$ having a lower dispersant demand for alkaline dispersant) the gloss could be improved through the additional use of AMP-95 indicating an excellent dispersion and film forming process (Table 2).

Further rheological data for a freesheet topcoating colour at 61.7% solids are shown in Figures 12 and 13.

6 Example of the Use of AMP-95™ in a Paper Coating Application – Post Added

Figure 14 shows the effect when AMP-95™ was post added to an already formed coating colour in response to an in-mill problem.

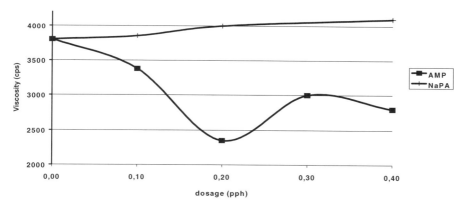

Figure 12 *Dispersant demand at low shear rate (Brookfield 100 rpm)*

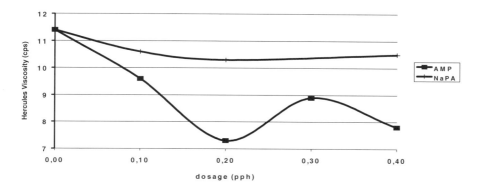

Figure 13 *Dispersant demand at high shear rate (Hercules, bob E, 4400 rpm)*

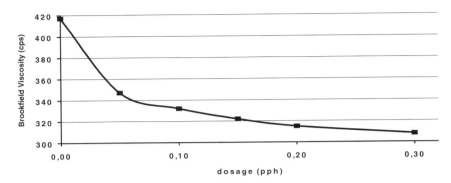

Figure 14 *AMP added to actual mill coating to solve problem*

7 Improved Rheology

AMP-95™ in paper coatings not only improves the rheology and gloss; it also has some side effects, which further enhance the coating performance. Due to its boiling point being higher than water, AMP-95™ ensures stable pH control, alkalinity and controlled solvent evaporation during the drying process.

8 Summary

AMP-95™ is liquid and can easily be dosed and incorporated into any coating formulation. The use level is typically between 0.15% and 0.25% based on solid content of the formulation. It is recommend using at least a part of AMP already during the grinding step to take full advantage of all its functionality. One typically will notice an improved wetting and easier filler incorporation with lower viscosities. In fact, because of the high performance, the formulator has to watch the viscosity carefully and has to maintain a viscosity level high enough to ensure sufficient energy input (shear force) for grinding. Adjusting the water content for the grind easily does this. In lots of cases AMP-95™ was used as a post add and also improved performance was found.

Through the use of the additive AMP-95™ several physical properties of the paper coating can be influenced and improved. Not only an ideal dispersion, viscosity and rheology profile can be achieved, due to the use of AMP-95™ also improved runnability, less dispersion defects, improved coverage and gloss was found. Hand in hand with this goes significantly reduced defects as streaks, scratches and mottling. In a lot of cases the use of AMP-95™ enables the formulator to achieve higher pigment loading or a safer production run due to widening the application window with the same % solids.

Pigment Dispersion Technology for the Paper Industry

D.J. Mogridge,[1] J.S. Phipps,[1] K.R. Rogan[1] and D.R. Skuse[2]

[1]IMERYS MINERALS LTD, JOHN KEAY HOUSE, ST. AUSTELL, CORNWALL, PL25 4DJ, UK
[2]IMERYS MINERALS INC., PO BOX 471, 618 KAOLIN ROAD, SANDERSVILLE, GA, 31082, USA

1 Introduction

Concentrated aqueous dispersions of fine, particulate kaolin and calcium carbonate are used extensively in the coating of paper. Currently more than 10 million tonnes of such dispersions are used annually in North America and Europe alone, and worldwide consumption is growing at around 3% per year.

Generally, paper is coated to enhance its visual appearance. Firstly, coatings improve the optical properties of the blank sheet, *i.e.* its brightness (or whiteness), opacity and gloss. Secondly, they improve the physical properties of the sheet that will influence the quality of the final printed image, such as surface smoothness, porosity and ink absorption. In order to achieve these aims economically, paper coatings must be applied at high speeds and with minimal drying costs.

Kaolin and calcium carbonate pigments are the major components of most paper coating formulations. Particle size distributions are chosen to maximise the optical benefits that they confer and to allow the production of formulations that can be easily pumped, handled and coated. Coating pigments are generally produced and transported as slurries, and so it is important to obtain dispersions with the highest possible solids content, both to minimise drying costs and the quantity of water transported. This paper examines the dispersant technology that is currently used in the industry and the mechanisms by which it functions, and illustrates ways in which this technology has developed and could be further improved in the future.

2 Paper Coating Technology Today

Figure 1 is a schematic diagram of a typical blade coater. This is still the most common form of coating device used in the paper industry, although new

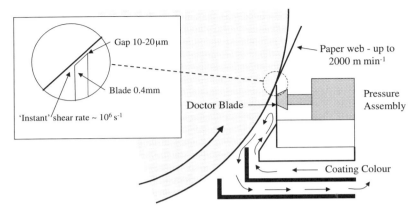

Figure 1 *Schematic diagram of blade coater*

technologies such as jet coaters and metered roll applicators are becoming more popular and may in time supersede it. A continuous web of paper passes over a large rotating drum, and is brought into close contact with a coating 'head'. The coating formulation, known in the industry as a coating 'colour', is circulated around the head and makes contact with the moving web of paper over a small region a few centimetres in length. At the end of this region a doctor blade is pressed against the paper, so that the desired coating thickness can be controlled by the pressure applied to the blade. A typical doctor blade is around 0.4 mm thick, and the gap between it and the moving paper web is of the order of $10{-}20\,\mu m$. At coating speeds of up to 2000 m min^{-1}, this translates to a shear rate in the order of $10^6\,s^{-1}$ over a period of $10\,\mu s$.

Laboratory measurements of shear rates of this magnitude are not easily accessible, but the behaviour of a formulation at a shear rate of the order of $10^4\,s^{-1}$ will usually indicate whether or not a coating colour will run successfully.

The optimum particle size for light scattering is of the order of the wavelength of light used and so most coating pigments have a mean particle size in the $0.3{-}0.8\,\mu m$ range. They follow approximately log/normal size distributions, with a standard deviation of $0.3{-}0.6$, which means that a typical coating pigment will have 95% by mass of its particles less than $2\,\mu m$ in diameter, 50% less than $0.5\,\mu m$ and 25% less than $0.25\,\mu m$. Some 80–95% of the solids component of a coating colour will be pigment, with 5–15% latex or starch binder and minor quantities of other components such as low shear-rate thickeners, lubricants and fluorescent whitening agents. The papermaker will want to run the coating colour at the highest possible solids, which can be as high as 65% by mass where the pigment is primarily calcium carbonate. Coating colours are highly shear-thinning materials, with apparent viscosities in the region of 1000 mPas at $10\,s^{-1}$ and 50 mPas at $10^4\,s^{-1}$.

The normal dispersants used for both kaolin and calcium carbonate pigments are aqueous solutions of sodium polyacrylate. These are prepared by free radical polymerisation using various combinations of initiators and terminators which may be proprietary to the manufacturer. Number average molecular weights are

generally less than 10 000. The dispersant is added at an addition rate of up to
1% by mass of the pigment in order to obtain a Brookfield 100 rpm apparent
viscosity (which corresponds to a maximum shear rate of approximately 20^{-1})[1]
of 500 mPas or less. Figure 2 and Figure 3 show some typical viscosity *vs.* dose
curves for calcium carbonate and kaolin slurries respectively. The values T_0, T_1
and T_{24} refer to the viscosities as measured immediately, one hour and twenty-
four hours after slurry preparation. Note that the optimum dose for kaolin is

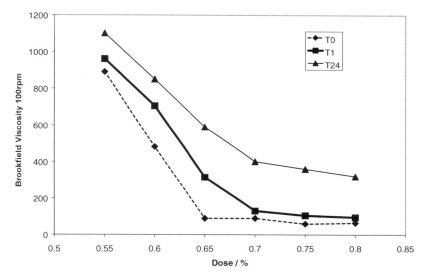

Figure 2 *Typical viscosity* vs. *dispersant dose for Na polyacrylate on coating grade calcium
carbonate (0.5 μm – 76% solids)*

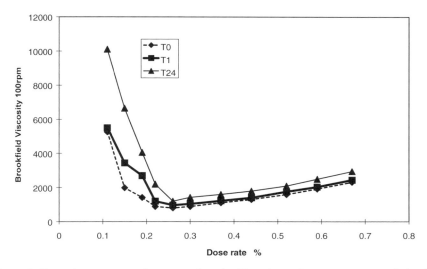

Figure 3 *Typical viscosity* vs. *dispersant dose for Na polyacrylate on coating grade kaolin
(0.6 μm – 68% solids)*

substantially lower than that for calcium carbonate. This is believed to be because the dispersant only adsorbs on the positively-charged edges of the plate-like kaolin particles; the faces are already negatively charged over a wide pH range.

3 Colloid Chemistry of Sodium Polyacrylate Dispersed Calcium Carbonate Slurries

In this section we give a description of the colloidal interactions which govern the state of dispersion and rheology of calcium carbonate slurries. The description is based on a series of fundamental experimental studies that have been described in more detail in previous publications.[2,3]

3.1 Experimental Details

A series of slurries of high solids, dispersed ground calcium carbonate were prepared. Firstly, a sample of Carrara marble was ground without dispersant to a mean size of 0.5 μm. This was then filtered to a solids level of 76% (volume fraction 46%), before being thoroughly mixed with the required dose of sodium polyacrylate dispersant and adjusted to a constant solids level of 70%. Doses ranged from zero to 25 mg g^{-1} (2.5%) of calcium carbonate. Samples were then left for two days to reach a steady-state adsorption level. After this they were sheared and their viscosity (Brookfield RV, 100 rpm, spindle 3) was measured.

The amount of polymer adsorbed on each sample was measured by pressure filtration through a 0.1 μm filter, followed by analysis of the filtrate for residual polymer by gel permeation chromatography with refractive index determination. Particle zeta potentials were measured by taking a small sample of the solids from the centrifuge and re-suspending them in the supernatant prior to analysis in a Malvern Instruments 'Zetasizer'. The concentration of all other types of ions in the supernatant was analysed by ICP atomic emission spectroscopy.

3.2 Results

Figure 4 shows the viscosity of the samples as a function of dispersant dose. Note that the viscosity decreases more or less monotonically with dose throughout and beyond the normal commercial dosage range, but that the reduction in viscosity at doses above 1% is relatively small.

Figure 5 shows the adsorption isotherm as calculated from the dose levels and the measurements of residual polymer in solution. The isotherm is of the high affinity type and can be divided into three regions. In region 1, at low polymer doses, almost all of the polymer added is adsorbed onto the particle surfaces and there is no detectable polymer in solution. In region 2, at intermediate doses, there is a small rise in the amount of polymer adsorbed but the majority of the extra polymer added in this region stays in solution. In region 3, at high polymer

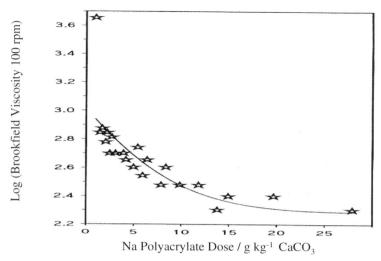

Figure 4 *Log (viscosity)* vs. *Dose for Na polyacrylate on 0.5 μm calcium carbonate at 70% solids*

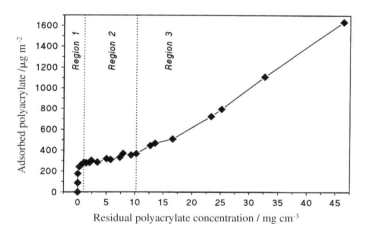

Figure 5 *Adsorption isotherm of Na polyacrylate on 0.5 μm calcium carbonate at 70% solids*

doses, the adsorbed amount rises again more steeply, whilst the solution concentration continues to rise substantially.

Figure 6 shows the total ionic strength, calculated from the analysis of all ions in solution including the polyacrylate, as a function of polymer dose. Apart from a small deviation in region 1, the ionic strength rises almost linearly with polymer dose over the whole range. Ionic strength rises less rapidly in region 1 because the added polymer all adsorbs on the surface, and so the increase in ionic strength is caused only by release of sodium and carbonate/bicarbonate ions into solution.

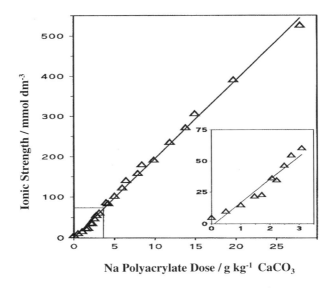

Figure 6 *Effect of Na polyacrylate dose on solution ionic strength*

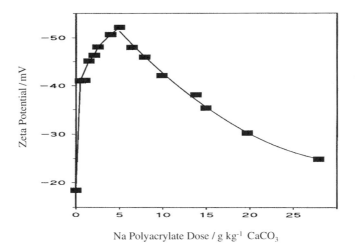

Figure 7 *Zeta potential of calcium carbonate as a function of Na polyacrylate dose*

Even at a dose level of $5\,mg\,g^{-1}$ some 80% of the contribution to ionic strength comes from unadsorbed sodium polyacrylate.

Figure 7 shows the effect of polymer dose on the measured zeta potential. Zeta potential continues to rise quite steeply through regions 1 and 2, but then decreases rapidly in region 3. This sharp decrease could be a result of the reduction in ionisation of the polymer as the adsorption density rises; however, recently published studies[4] on tethered polyacrylate chains at various surface densities suggest that at the pH of these experiments (8.5–9.5) the polymer should remain fully ionised despite the proximity of the charges to each other. The decrease in zeta potential is thus more likely to be caused by a shift in the position of the slip plane as the particles move or by specific ion binding to the adsorbed layer.

4 Discussion

Zeta potential and ionic strength data were used to calculate the interparticle repulsion according to the Derjaguin–Landau–Verwey–Overbeek (DLVO) theory. According to these calculations, a maximum in the interparticle repulsion energy of around 120 kT should occur at the transition between regions 1 and 2, and thereafter the repulsion should fall rapidly, reaching zero at the end of region 2 and becoming attractive throughout region 3. Clearly this is not consistent with the viscosity data which suggest that the interparticle repulsion continues to rise through both regions 2 and 3. This indicates that the repulsion must also have a steric component which becomes increasingly important as the polymer dose is raised.

In order to be able to include a steric contribution in the interparticle energy calculation, an estimate of the adsorbed layer thickness is required. This is very difficult to access experimentally; probably the only technique which might be able to provide an estimate is small-angle neutron scattering which was beyond the scope of this work. As a result, a theoretical estimation of the thickness was made, based on a few key observations. This is described below.

Figure 8 shows the radius of gyration (R_g) of sodium polyacrylate in solution as a function of ionic strength, calculated from intrinsic viscosity measurements.[5] It can be seen that R_g varies almost linearly with the inverse square root of the ionic strength (I). By fitting the data to a simple second-order polynomial, the value of R_g in solution was interpolated across the whole of the ionic strength range studied. Likewise, the area per molecule or molecular 'footprint' of the adsorbed polymer in regions 1 and 2 was calculated from the adsorbed amount and surface area of the sample and this was then converted to the apparent R_g of the adsorbed polymer by multiplying by a simple geometric conversion factor. This is also shown in Figure 8. As with the solution polymer, the of R_g the adsorbed polymer in a completed monolayer (region 2) varies almost linearly with $I^{-1/2}$. The R_g data for the adsorbed polymer in the completed monolayer was fitted to a polynomial of equal gradient to that fitted to the data for the solution polymer, and this was then extrapolated into region 3 in order to obtain a value for the R_g of the polymer in the region of multilayer adsorption. It was assumed that in this region all polymer molecules in completed monolayers have the same R_g regardless of which layer they are in. The R_g of the adsorbed polymer in a completed monolayer is lower than the corresponding R_g in solution because the polymer chains are laterally compressed by the presence of neighbouring polymer molecules. The assumption was also made that the polymer occupies the same volume on the surface as in the bulk at each value of ionic strength, so

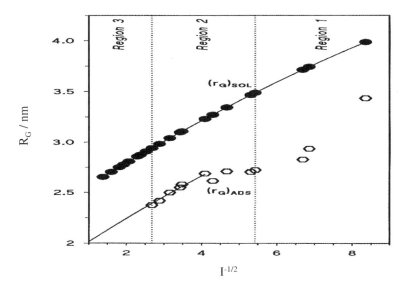

Figure 8 *Radius of gyration of Na polyacrylate in solution and adsorbed on calcium carbonate surface as a function of ionic strength*

that laterally-compressed adsorbed polymer chains occupy a cylindrical volume, the length of which (and thus the thickness of the adsorbed monolayer) can be calculated from the polymer volume and the value of R_g.

The thickness of the adsorbed polymer layer at each dose level was thus estimated from the following sequence of steps:

1. The number of polymer molecules adsorbed per unit area was calculated from adsorption isotherm data and the surface area per unit mass of the sample.
2. The radius of gyration and area per molecule in a completed monolayer was determined from the polynomial fit of adsorbed R_g vs. I.
3. The number of molecules in a completed monolayer was then calculated and hence also the number of completed layers and the number of molecules in the outer, incomplete layer.
4. The area per molecule in the outer, incomplete layer was calculated.
5. By equating the polymer volume in solution with the polymer volume on the surface, the thickness of each layer was calculated and these values were summed to obtain the total layer thickness.

This is shown schematically in Figure 9.

The total layer thickness calculated using this approach is shown in Figure 10. The most notable feature is the substantial contraction of the layer thickness in region 2. This occurs because in this region the adsorbed amount remains almost constant, whilst the solution ionic strength rises rapidly. The contraction of the layer in this region implies that both the steric and electrostatic components of the interparticle repulsion are reduced by increasing the solution ionic strength. In region 3, where multilayers are forming, the effect of increasing ionic strength

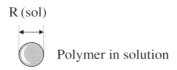

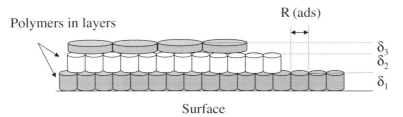

Figure 9 *Schematic representation of model used for estimation of polymer layer thickness*

is offset by the increasing amount of polymer adsorbing onto the surface, and hence the layer thickness increases.

Using this data for the layer thickness, the total energy of interaction was calculated by summing the electrostatic and steric contributions, the latter of which was calculated according to the method of Ottewill.[6] Total interaction energies at three values of interparticle separation are shown in Figure 11 as a function of polymer dose.

From Figure 11 it can be seen that the inclusion of a steric term changes the calculated interparticle repulsion substantially. Whereas the electrostatic-only

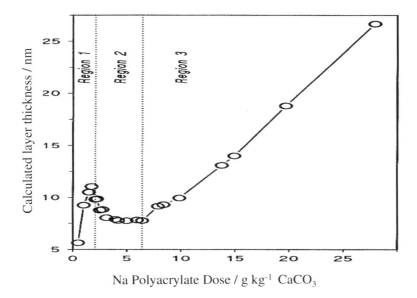

Figure 10 *Calculated layer thickness as a function of polymer addition level*

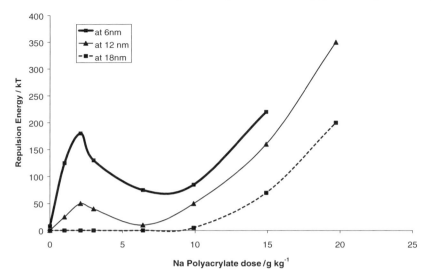

Figure 11 *Calculated total interaction energies at fixed interparticle separation as a function of polymer dose*

calculation predicted that the maximum repulsion would have shrunk to zero at a dose level of $10 \, \text{mg g}^{-1}$, the combined calculation predicts that close to the surface it remains above $100 \, \text{kT}$ at all doses, and rises substantially at the onset of multilayer formation. Hence by including a steric term the calculation agrees with the observation that the slurry remains dispersed at all doses above $1 \, \text{mg g}^{-1}$. However, the model also predicts that the repulsion should reach a minimum at a dose of around $8 \, \text{mg g}^{-1}$, and this is clearly not indicated by the viscosity data either for this sample or for paper-coating grade calcium carbonates in general.

There are a number of assumptions made in the model that are questionable, and these are probably responsible for the discrepancy between the predicted and observed behaviour of the slurries. For example, the steric contribution has been calculated assuming that the adsorbed layer has a well-defined thickness. For adsorbed polymers this is unlikely to be the case, as the volume fraction profile of the polymer will decay gradually as a function of distance from the surface. Furthermore, it was assumed that the effective ionic strength in the adsorbed polymer layer is the same as in solution. However, this also is unlikely since one of the main components of the solution ionic strength is the polymer itself, and unadsorbed polymer will be excluded from the adsorbed layer. Finally, the connectivity of the charged groups on the polymer was not considered, so its contribution to ionic strength may have been overestimated.

Despite these misgivings, the model is useful in that it highlights a number of important points. Firstly, it confirms that steric repulsion is the main driving force for dispersion at the higher polymer doses used, and thus explains the continued performance of the polymer as a dispersant even at relatively high values of ionic strength. Secondly, it emphasises that increasing ionic strength is harmful not just to the electrostatic component of the repulsion, but also to the steric component because it compresses the adsorbed layer. As a result, it shows that to improve the effectiveness of the dispersant measures should be taken which maximise the amount of polymer on the particle surface and the layer thickness, and which minimise the solution ionic strength.

5 Application of Model Conclusions to Slurry Dispersant Optimisation

From the preceding discussion, the most obvious way to improve the performance of polyacrylate-dispersed calcium carbonate slurries is to reduce the solution ionic strength by reducing the amount of non-adsorbed polymer in solution.

An efficient way of increasing the amount of polymer adsorbed whilst minimising the amount in solution and hence minimising ionic strength is to increase the affinity of the polymer for the particle surface. For polyacrylate on calcium carbonate, this can be done quite simply by changing the degree of neutralisation of the added polymer,[7] as any acid groups will react directly with the particle surface and thus bind strongly to it:

$$2R\text{-}CO_2H + CaCO_3 \rightarrow (R\text{-}CO_2)_2Ca + CO_2$$

Furthermore, binding of the polyacrylate in this way does not lead to the release of sodium ions into solution as it does when the 100% sodium neutralised polymer is used. The final equilibrium slurry pH remains around 8.5 because it is buffered by the carbonate.

Although this approach does lead to stronger binding of the polymer, it will also reduce the amount of free charges on the adsorbed chains and most likely lead to a flatter conformation. As can be seen from Figure 12, as a result of these two conflicting effects there is an optimum range of initial neutralisation that leads to low slurry viscosities.

A similar situation is seen for kaolin, although in this case it is the final slurry pH which dictates the dispersant performance rather than the initial state of the polymer, since the particles do not buffer the solution in the same way as calcium carbonate. As pH is decreased the edges of the kaolin particles become more positvely charged which leads to stronger adsorption of the negatively charged polymer. However, if the pH gets too low the degree of ionisation of the polymer decreases which weakens the adsorption. Likewise at high pH the polymer is highly negatively charged, but as the positive charge on the particles also reduces again the affinity of the surface for the polymer is reduced. As can be seen in Figure 13, there is an optimum pH range of 6–8, outside of which the viscosity, particularly after standing, begins to rise.

One final way in which to reduce the amount of polymer in solution is to use a narrow molecular weight distribution. It is well known that narrow distributions give sharper adsorption isotherms. Commercial polymers are made by free

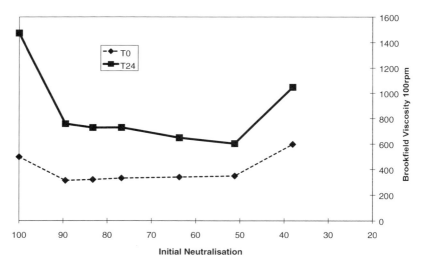

Figure 12 *Effect of degree of neutralisation of polyacrylic acid on viscosity of dispersed calcium carbonate (0.5 μm, 76% solids)*

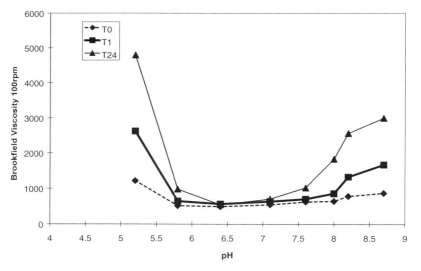

Figure 13 *Effect of pH on viscosity of polyacrylate dispersed kaolin (0.6 μm, 70% solids)*

radical polymerisation methods that do not naturally lead to narrow distributions, but sample polydispersity can be reduced by molecular weight fractionation techniques. Figure 14 shows the effect of such a fractionation on the viscosity of a calcium carbonate slurry.

6 Summary

The paper industry requires high solids dispersions of kaolin and calcium

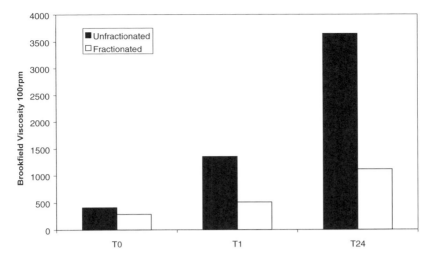

Figure 14 *Effect of polymer M_w fractionation on viscosity of dispersed calcium carbonate (0.5 µm, 76% solids)*

carbonate with good rheology for both cost and quality reasons. Sodium polyacrylate is the most widely used dispersant for this purpose because it is both cheap and effective.

Fundamental studies of the effects of sodium polyacrylate have shown it to be an 'electrosteric' dispersant, for which the steric component of interparticle repulsion is the dominant one at the typical dose rates used commercially. In such dispersions the main contributor to solution ionic strength is unadsorbed polyacrylate. As well as representing a waste of valuable dispersant polymer, this unadsorbed material also reduces the effectiveness of the adsorbed material as a steric dispersant by compressing the conformation of the adsorbed layer.

Measures taken to minimise the amount of unadsorbed polymer in solution have been shown to improve the viscosity performance of both kaolin and calcium carbonate slurries. These include adjustment of the pH or initial neutralisation of the polymer to increase its affinity to the surface and fractionation of the polymer to reduce the width of its molecular weight distribution.

References

1. H.A. Barnes, *Applied Rheology*, 2001, **11**(2), 89.
2. K.R. Rogan, A.C. Bentham, G.W.A. Beard, I.A. George and D.R. Skuse, *Progress in Colloid and Polymer Science*, 1994, **97**, 97.
3. K.R. Rogan, A.C. Bentham, I.A. George and D.R. Skuse, *Colloid and Polymer Science*, 1994, **272**, 1175.
4. E.P.K. Currie, N.J. Avena and M.A. Cohen Stuart, in *'Polyelectrolytes: Proceedings of the 50th Yamada Conference'*, I. Nada and E. Kokufuta (eds.), Yamada Science Foundation, 1999, 384.

5. K.R. Rogan, *Colloid and Polymer Science*, 1995, **273**, 364.
6. R.H. Ottewill, in *'Scientific Methods for the Study of Polymer Colloids and Their Applications'*, F. Candau and R.H. Ottewill (eds.), NATO ASI Series, Kluwer Academic, Dordrecht, 1990, 129.
7. O. Gonnet, G. Ravet and J. Rousset, *US PAT.*, 4840985.

Selective Processing

An AFM and XPS Investigation of the Selective Flocculation of Kaolinite from a Mineral Mixture

J.S. Dalton,[1] G.C. Allen,[1] K.R. Hallam,[1] N.J. Elton,[2] J.J. Hooper[3] and D.R. Skuse[4]

[1]INTERFACE ANALYSIS CENTRE, UNIVERSITY OF BRISTOL, 121 ST. MICHAELS HILL, BRISTOL, BS2 8BS, UK
[2]SCHOOL OF ENGINEERING & COMPUTER SCIENCE, UNIVERSITY OF EXETER, NORTH PARK ROAD, EXETER, DEVON, EX4 4QF, UK
[3]IMERYS MINERALS LTD, PAR MOOR LABS, PAR MOOR ROAD, ST. AUSTELL, CORNWALL, PL24 2SQ, UK
[4]IMERYS MINERALS INC., PO BOX 471, 618 KAOLIN ROAD, SANDERSVILLE, GA, 31082, USA

1 Introduction

Mineral segregation in industry relies heavily on the selective adsorption of macromolecules onto the surfaces of those minerals that have particular industrial applications. This selectivity is governed mainly by the surface chemistry of the mineral and the type of polymer used as a flocculant.[1,2] Effectiveness of flocculation depends upon the charge, concentration and molecular weight of the polymer, and also the pH and salt concentration of the clay suspension. The bonding between the anionic flocculant polyacrylamide (PAM) and clay mineral surfaces has been effectively reviewed recently by Hocking *et al.*[3] and the reader is referred to this should they require an in-depth literature review. For more information on general colloidal chemistry of clay suspensions the reader is referred to the review of Luckham and Rossi.[4]

There is general agreement that the main mechanism for the adsorption of cationic polymers onto clay surfaces occurs by charge neutralisation.[5] There is less agreement on the adsorption of anionic polymers, such as hydrolysed polyacrylamide (HPAM), onto mineral surfaces. It has been suggested that this occurs by cation bridging whereby divalent ions bridge negatively charged sites on the clay and carboxyl groups on the polymer.[5,6] However, work by Peng and Di has shown that Ca^{2+} and Al^{3+} actually inhibit kaolin flocculation as these

71

ions decrease the charge density and extendibility of the polymer.[7] Work of Pefferforn and Nazbar showed that adsorption is governed by hydrogen bonding between the amide group on HPAM and hydroxyl groups at the surface of the clay.[8–11] This, in turn, depends on the pH of the mineral suspension.

A recent paper by Laird[12] investigated the efficacy of HPAM flocculation of kaolinite, illite and quartz by carrying out visible absorption experiments. He concluded that HPAM more effectively flocculates kaolinite than quartz or illite. This was also the conclusion of previous work by Allen *et al.* who studied the adsorption of HPAM onto kaolinite, quartz and feldspar at various HPAM concentrations and solution pH by X-ray photoelectron spectroscopy (XPS).[13–15] Much of the previous work on polyacrylamide adsorption onto aluminosilicates monitored the adsorbed amount by viscometry,[16] carbon analysis[17,18] and radiotracer techniques.[10] These methods rely on following adsorption by subtraction from that detected in solution.

This paper contributes to the literature by quantifying anionic polymer adsorption onto the clay minerals kaolinite, feldspar, mica and quartz by X-ray photoelectron spectroscopy (XPS). XPS measures the sorbed amount directly rather than by a subtraction technique. This enables an insight into how effective selective flocculation is for obtaining kaolinite from a mineral mixture. Atomic force microscopy (AFM) is also used to image polymer adsorption onto mineral surfaces and the effectiveness of this technique applied to mineral surfaces is discussed here.

2 Materials and Methods

2.1 Materials

2.1.1 Minerals. The polymer used in this study is partially hydrolysed polyacrylamide (HPAM), supplied by Allied Colloids Ltd, Low Moor, Bradford, West Yorkshire, England, UK. Approximately 10 mol% of the monomer units are hydrolysed and are present as sodium acrylate with a polyacrylamide-co-acrylate copolymer having a molecular mass of approximately 6×10^6 D. The study minerals quartz, kaolinite and feldspar were collected from the Lee Moor China Clay Pit in Devon, UK. Mica was obtained in sheet form from Agar Scientific.

Minerals were ground such that a size distribution ranging from sub-micron to millimetre particles were obtained. For the experiments described here, it is desirable to use monodisperse minerals. To this end, a sedimentation technique was used to obtain minerals in the particle size range (effective Stokes radius) (i) 10–20 μm; and (ii) above 20 μm.

2.1.2 Mineral Flocculation for XPS Analysis. A stock solution of 500 ppm HPAM in water was prepared and left to diffuse and condition for 1 hour due to the viscosity of the solution. 1 g of the mineral was made into a slurry with 10 cm^3 distilled water and this was then made up to 50 cm^3 in a measuring cylinder. The solution pH was then altered to 9.0 with the relevant amount of 0.1 M NaOH solution. An amount of 500 ppm HPAM was added to obtain the desired concentration and the cylinder is made up to 100 cm^3. The cylinder was shaken and rotated for 1 minute, and left to stand for a further 30 minutes by which time the mineral had settled by gravity. The supernatant HPAM solution was then decanted. The resulting mineral slurry was transferred to a conical flask and washed three times with distilled water, each time the supernatent being decanted off. This rinsing process ensured any unadsorbed or very weakly adsorbed HPAM was removed. The remaining slurry was deposited on a silica support and left to dry for 24 hours, ready for XPS investigation.

For imaging XPS, this process was carried out on a kaolinite and coarse quartz (above 20 μm stokes radius) 1:1 mix.

2.1.3 Mineral Preparation for AFM Imaging. Both untreated and polymer treated mineral samples were attached to metallic stubs using a rapid acting araldite from Ciba Speciality Chemicals. The araldite permitted careful placing of the minerals such that macroscopically flat areas were accessible for AFM analysis.

2.2 Methods

2.2.1 X-ray Photoelectron Spectroscopy (XPS). XPS analyses the near-surface of materials with analysis depths up to 5 nm. It can provide quantitative chemical information as well as oxidation and structural environments on all elements apart from hydrogen and helium.[19] XPS spectra were acquired using a VG Scientific Escascope spectrometer using Mg K_α X-rays (300 W; 15 kV, 20 mA) and a hemispherical electron energy analyser. Data acquisition and manipulation were carried out using manufacturer-supplied VGS5250 software. During acquisition of spectra, the pressure in the main chamber was maintained at 8×10^{-9} Torr to ensure a clean sample surface. Charge referencing was carried out against adventitious hydrocarbon ($C\,1s = 284.8$ eV)[19] from pump oil contamination. Imaging XPS was carried out in constant retard ratio mode with a 1200 μm field-of-view giving <50 μm resolution. A 500 eV flood gun was used to locate the region of interest. Spectra were then acquired in this region to obtain specific photoelectron peaks and backgrounds. The background is a few electron-volts lower in energy than the photoelectron peak where the count intensity is purely due to the averaged effect of inelastically scattered electrons. A resultant image of the background subtracted from peak map can help in reducing the effects of surface topography.

2.2.2 Atomic Force Microscopy (AFM). The AFM[20,21] images were obtained on a Digital Instruments Dimension 3100 microscope which was mounted on an air cushioned metal table to reduce external vibrations. Images were collected and processed using Nanoscope version 4.42rl software. The instrument was operated in tapping mode (TMAFM)[22,23] in air using etched silicon cantilevers with a spring constant of 50 N m^{-1}. Tapping mode was used to minimise the lateral forces the tip exerted on the sample surface which could disrupt and deform the polymer layer. A TMAFM scan allowed the acquisition of a height image and the corresponding phase image simultaneously. The phase contrast gave information on the viscoelasticity (material contrast) helping to distinguish between features of the hard substrate surface and the adsorbed soft polymer.[24] A high setpoint, ensuring intermittent contact, and a slow scan rate (0.50 Hz) were chosen.

2.2.3 Scanning Electron Microscopy (SEM).   The SEM used was a commercial Hitachi S2300 instrument with a tungsten hairpin filament. An accelerating voltage of 25 keV was used and samples were gold coated to eliminate charging.

3 Results and Discussion

3.1 Monodispersed Minerals

SEM images of the monodispersed minerals are shown in Figure 1. Unfortunately it was not possible to obtain kaolinite particles in the desired size range (10–20 μm) due to flocculation occuring in the sedimentation process. It was not desirable to chemically treat the surface of the kaolinite to prevent this, as this would affect later results. To this end, the kaolinite was used untreated, as received, with the smaller size fraction shown below. Monodispersed minerals will allow direct comparison in HPAM adsorption isotherms.

(a)   (b)

(c)   (d)

Figure 1 (a) *Mica (50 μm marker); (b) feldspar (100 μm marker); (c) quartz (50 μm marker); all monodisperse with effective Stokes radius 10–20 μm. (d) Kaolinite (5 μm marker)*

3.2 XPS: HPAM Adsorption Isotherms

3.2.1 Minerals. X-ray photoelectron spectra were run on the untreated minerals to obtain peak binding energies and atomic percent in the surface. This is presented in Table 1. It should be noted that the surface carbon (C *1s*) is due to adventitious hydrocarbon impurity from the UHV diffusion pump. The nitrogen impurity present on the surface of the minerals was organic, probably amine or amide, and is thought to be associated with the origin of the minerals. Many organic molecules, for example pesticides, are seen to sequester into weathering-produced 'dead-end' pores at mineral surfaces.[24] The XPS spectrum of a HPAM film was consistent with its chemical structure.

3.2.2 Minerals Treated with Polyacrylamide. Minerals were treated with HPAM as outlined in Section 2. Any unadsorbed or weakly bound HPAM was removed from the mineral by washing and rinsing. Adsorption on the mineral surface was monitored by the N *1s* photoelectron peak from the HPAM. To compare HPAM adsorption between minerals, peak integration was used to obtain surface atomic % of N *1s*, Si *2p* and Al *2p*. The ratio N*1s*:(Si*2p* + Al*2p*) was used to normalise all the samples, which eliminates the contribution of C *1s* hydrocarbon contamination from the analysis.

Multiple XPS scans gave error values as $\pm 5\%$ on atomic surface percent and $\pm 0.2\,\text{eV}$ on binding energy.

Figure 2(a) shows how the adsorbed amount of HPAM on kaolinite increased with bulk concentration HPAM. For clarity, the 200 ppm measurement was omitted, but the peak intensity was not seen to rise above 50 ppm. It would appear that there is surface saturation above this 50 ppm threshold limit. Figure 2(b) shows an HPAM adsorption isotherm for both kaolinite and feldspar. Neither the quartz or mica showed any adsorbed HPAM up to a bulk concentration of 200 ppm. The N *1s* atomic percent did not vary from the background adsorption at 0 ppm, at least within the error margin in these experiments.

Table 1 *Peak binding energies and surface atomic percent for the minerals*

	Si 2p	Al 2p	O 1s	K 2p	N 1s	C 1s
kaolinite						
binding energy/eV	102.8	74.4	532.2		399.8	294.8
atomic %	9.2	6.7	47.1		0.4	36.6
quartz						
binding energy/eV	102.8	74.4	532.4		399.8	284.8
atomic %	20.5	1.5	49.5		0.7	27.7
mica						
binding energy/eV	102.5	74.2	531.8	292.9	399.8	284.8
atomic %	13.6	9.7	57.3	4.9	0.3	14.3
feldspar						
binding energy/eV	102.5	74.2	531.8	292.9	399.8	284.8
atomic %	15.7	3.5	49.8	4.4	0.2	26.4

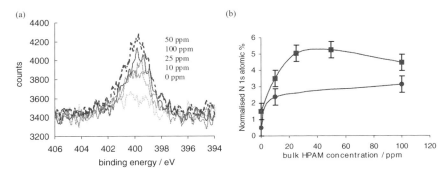

Figure 2 (*a*) *Using the* N 1s *photoelectron peak to follow the adsorption of HPAM on kaolinite. Inset shows bulk HPAM concentration in order of peak intensity.* (*b*) *HPAM adsorption isotherm onto kaolinite* (*squares*) *and feldspar* (*circles*)

3.3 XPS Imaging: Selective Adsorption of HPAM in Mineral Mixtures

In order to monitor the selective adsorption of HPAM onto one mineral in preference to another, imaging XPS was carried out on a mixture of kaolinite and quartz. Figure 3 shows the backscattered electron image of the kaolin–quartz matrix in addition to the Si *2p* and Al *2p* photoelectron images. The field of view is 1200 μm. Both (b) and (c) are resultant photoelectron images, the background having been subtracted from the peak image. Using this technique it is possible to differentiate the larger quartz crystals embedded in the kaolin matrix due to the presence and absence of Si and Al respectively. Note, however, there is a signal from Al *2p* on the surface of the quartz crystal.

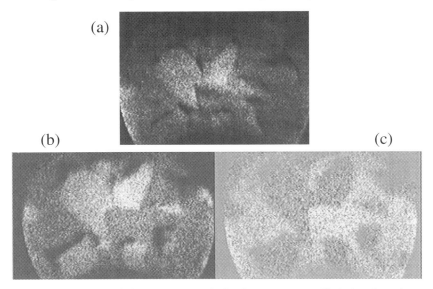

Figure 3 (*a*) *Backscattered electron image of a kaolin–quartz mix;* (*b*) *Si* 2p *photoelectron peak image;* (*c*) *Al* 2p *photoelectron peak image*

It was attempted to gain N *1s* images to identify where the HPAM adsorbed, but images were generally poor. Due to the topography of the sample there was a large amount of shadowing. Photoelectrons are not emitted from this region giving poor images. It appears that when the kaolinite flocculates, it entraps other minerals within the flocs, hence minerals such as quartz will have a covering of kaolinite on their surface. This is seen in the SEM images in Figure 4. Therefore the N *1s* map will represent the HPAM adsorbed on the kaolinite covering the quartz.

XPS measurements on mineral mixtures of this kind proved an ineffective means of monitoring HPAM adsorption.

3.4 AFM Analysis of HPAM Adsorbed to Kaolinite

A review on AFM imaging of mineral surfaces has been previously given by Maurice.[25] In order to effectively image minerals in the micron size range it is essential that the surface is relatively flat,[26] at least less than 5 μm roughness, otherwise image quality is poor. AFM is capable of imaging both large and small areas of mineral surfaces and this is demonstrated in Figure 5.

Figure 6 shows a typical AFM topographical (a) and phase (b) image of untreated kaolinite. The phase image shows no contrast on the plates, which confirms a clean surface. The hexagonal nature of the plates is seen clearly in both images. The HPAM treated kaolinite is also shown in (c) and (d). Initially concentrating on the topological image (c) it can be seen that these plates are less distinguished than the untreated and this was due to adsorbed HPAM. The phase contrast (d) occurs mainly from topographical polymer layer effects indicating that the adsorption on the basal plane is at least of monolayer coverage over most of the plate. In addition the characteristic 'booksheets' on the lateral faces of kaolinite are also seen. The adsorption to the basal surfaces of the kaolinite is generally deemed to be due to hydrophobic bonding and also hydrogen bonding to the aluminol groups.

Figure 4 *SEM images of a kaolinite–quartz–mica mineral mix after flocculation by HPAM.*
The kaolin covers other minerals 'encapsulating' them inside the flocs

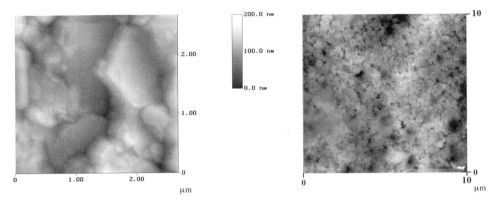

Figure 5 *AFM imaging of kaolinite*

Figure 6 *Topographic and phase imaging of untreated and HPAM treated kaolinite*

3.5 Adsorption Mechanisms

It is seen here that HPAM strongly adsorbs to kaolinite, causing flocculation, and also has affinity to the feldspar surface. Little or no adsorption was observed on the quartz or mica surface. Any adsorption mechanism must account for this.

The schematic in Figure 7 shows the suggested HPAM bonding mechanism to the kaolinite surface at low and high pH at both the siloxane and aluminol sites. Previous IR measurements have shown that interactions between polymer amide groups and oxygens of siloxanes were found to be very weak.[27] In addition, at pH 9.0 there will be repulsion between the carboxylate groups on the HPAM and the exposed $>$ Si–O$^-$ group. This explains little bonding to the quartz. Lee *et al.* measured the adsorption of polyacylamide onto gibbsite and observed a relatively high surface coverage.[18] It would appear that the strongest hydrogen bonding occurs at the aluminol sites on these minerals.

4 Conclusion

XPS can be used to quantify HPAM adsorption onto minerals at various polymer bulk concentrations. It is seen here that kaolinite has twice the affinity for HPAM than feldspar at pH 9.0 and 50 ppm. Little or no adsorption was monitored on the surface of quartz or mica. Imaging XPS to monitor selective adsorption of mineral mixes proved difficult. Flocculating a mineral mixture of kaolinite, mica and quartz caused the kaolin flocs to encapsulate the other minerals. This created a layer of kaolin on the quartz and mica prohibiting polymer mapping on their surfaces. It is shown that the effectiveness of the kaolin recovery is more strongly affected by encapsulation of other minerals during flocculation rather than the selective adsorption process.

Providing a flat surface can be achieved, tapping mode AFM topological and phase imaging can be used to investigate HPAM adsorption and mineral surfaces. Information on the film height, the strength of interaction between polymer–mineral surface and an indication of the adsorption process may be obtained. HPAM adsorption on kaolinite was observed to form a flat uniform layer

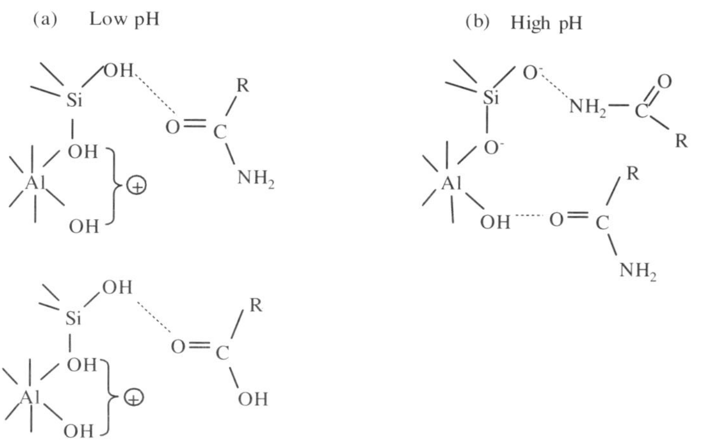

Figure 7 *Bonding mechanisms for HPAM onto kaolinite*

on the basal surface and is thought to occur *via* hydrogen bonding between the amide C=O group and the gibbsite Al–OH layer.

Acknowledgements

The authors would like to thank Mr Heiko Haschke and Mr Edward Leach for assistance with the AFM imaging.

References

1. G.J. Fleer, M.A. Cohen-Stuart, J.M.H.M. Scheutjens, T. Cosgrove and B. Vincent, *Polymers at Interfaces*, Chapman and Hall, London, 1993.
2. E. Dickinson and L. Eriksson, *Adv. Coll. Int. Sci.*, 1991, **34**, 1.
3. M.B. Hocking, K.A. Klimchuk and S. Lowen, *J.M.S.-Rev. Macromol. Chem. Phys.*, 1999, **C39**(2), 177.
4. P. Luckham and S. Rossi, *Adv. Coll. Int. Sci.*, 1999, **82**, 43.
5. S.M. Aly and J. Letey, *Soil Sci. Soc. Am.*, 1998, **52**, 1453.
6. V.C. Farmer, *Soil Sci.*, 1971, **112**, 62.
7. F.F. Peng and P. Di, *J. Coll. Int. Sci.*, 1994, **164**, 229.
8. L. Nabzar, E. Pefferkorn and R. Varoqui, *Coll. Surfaces*, 1988, **30**, 345.
9. L. Nabzar, E. Pefferkorn and R. Varoqui, *J. Coll. Int. Sci.*, 1984, **102**, 380.
10. E. Pefferkorn, L. Nabzar and A. Carroy, *J. Coll. Int. Sci.*, 1985, **106**, 94.
11. L. Nabzar and E. Pefferkorn, *J. Coll. Int. Sci.*, 1985, **108**, 243.
12. D.A. Laird, *Soil Sci.*, 1997, **162**, 826.
13. G.C. Allen, K.R. Hallam, J.R. Eastman, G.J. Graveling, V.K. Ragnarsdottir and D.R. Skuse, *Surf. Int. Anal.*, 1998, **26**, 518.
14. G.C. Allen, J.R. Eastman, K.R. Hallam, G.J. Graveling, V.K. Ragnarsdottir and D.R. Skuse, *Clay Minerals*, 1999, **34**, 51.
15. G.J. Graveling, K.V. Ragnarsdottir, G.C. Allen, J. Eastman, P.V. Brady, S.D. Balsey and D.R. Skuse, *Geochim. et Cosmochimica Acta*, 1997, **61**, 3515.
16. G. Girod, J.M. Lamarche and A. Foissey, *J. Coll. In. Sci.*, 1988, **12**, 265.
17. J. Lecourtier, L.T. Lee and G. Chauveteau, *Coll. Surfaces*, 1990, **47**, 219.
18. L.T. Lee, R. Rahbari, J. Lecourtier and G. Chauveteau, *J. Coll. Int. Sci.*, 1991, **147**, 351.
19. D. Briggs and M.P. Seah, *Practical Surface Analysis* (2nd edn.), Vol. 1, John Wiley, Chichester, 1992.
20. G. Binnig, C.F. Quate and C. Gerber, *Phys. Rev. Letters*, 1986, **56**, 930.
21. R. Wiesendanger, *Scanning Probe Microscopy and Spectroscopy*, Cambridge University Press, Cambridge, 1994.
22. Q. Zhong, D. Inniss, K. Kjoller and V.B. Elings, *Surf. Sci.*, 1993, **290**, L688.
23. D. Anselmetti, M. Dreier, R. Luethi, T. Richmond, E. Meyer, J. Frommer and H.-J. Guentherodt, *J. Vac. Sci. Technol.*, 1994, **B12**, 1500.
24. R.P. Schartzenbach, P.M. Geschwend and D.M. Imboden, *Environmental Organic Chemistry*, Wiley-Interscience, New York, 1993.
25. P.A. Maurice, *Coll. Surfaces*, 1996, **107**, 57.
26. A.E. Blum, in *Scanning Probe Microscopy of Clay Minerals*, Nagy and Blum (eds.), Clay Min. Soc., London, 1994.
27. J.P. Burelbach, S.G. Bankoff and S.H. Davis, *J. Fluid Mech.*, 1988, **195**, 463.

The Nature of Adsorption Sites on Unrefined and Ball Milled Kaolin. A Diffuse Reflectance Infrared Fourier Transform Spectroscopic Study

C. Breen,[1] J. Illés,[1] J. Yarwood[1] and D.R. Skuse[2]

[1]MATERIALS RESEARCH INSTITUTE, SHEFFIELD HALLAM UNIVERSITY, SHEFFIELD, S1 1WB, UK
[2]IMERYS MINERALS INC., PO BOX 471, 618 KAOLIN ROAD, SANDERSVILLE, GA 31082, USA

1 Introduction

Mechanical milling of mineral samples is often utilised to modify surface and bulk properties such as particle size and surface area. However, other changes accompany the milling process such as structural alterations, complete or partial amorphisation of the material together with an increase in the surface energy and surface reactivity which may lead to enhanced chemical reactivity. The effect of grinding on kaolin has received much attention and the results reviewed in detail.[1,2] Dry grinding of kaolin is known to fracture the individual crystals leading to an increase in surface area. However, scanning electron microscopy suggests that this increase may be attentuated as the small particles aggregate even at short grinding times. Indeed, mechanochemical treatment for as little as 1 h can reduce the temperature at which the technologically important kaolin to mullite transformation occurs.[3]

A wide variety of instrumental techniques, including X-ray diffraction, thermal analysis, electron microscopy, MAS-NMR and infrared spectroscopy,[1-6] have been employed at different levels of complexity to investigate the effects of mechanochemical treatment on kaolin. Unfortunately, vibrational spectroscopy has only been used at a superficial level in the study of milled kaolin despite the considerable contribution that it has made to the understanding of the structure and reactivity of kaolin itself.

Thermal analysis, and to a lesser extent infrared spectroscopy, has shown that grinding kaolin reduces the number of structural hydroxyls and increases the amount of sorbed water. However, the infrared studies have been restricted to transmission studies using KBr discs at room temperature so little information concerning the nature and the thermal stability of the different types of sorbed water has emerged.[6]

In this study variable temperature diffuse reflectance infrared Fourier transform spectroscopy (VT-DRIFTS), supported by X-ray diffraction (XRD), scanning electron microscopy (SEM) together with thermogravimetric analysis (TGA) have been used to investigate the nature and thermal stability of water molecules associated with kaolin which has been milled for 3, 10 and 30 minutes at room temperature. The 1500 to 1700 cm^{-1} region of the DRIFTS spectra of the milled kaolins has been particularly informative identifying at least two separate water populations of differing thermal stability which are reflected in the OH stretching region (3000–3600 cm^{-1}).

2 Materials and Methods

The kaolin sample used was supplied by English China Clays International Ltd. (now IMERYS). X-ray diffraction data indicated the presence of a mica impurity (<5 wt%). Separate 2 g samples were milled in a Retsch MM2 ball mill grinder in sealed containers at 45 Hz for 3, 10 and 30 minutes and transferred to airtight containers. Diffuse reflectance Fourier transform infrared spectroscopy (DRIFTS) spectra were recorded using a Graseby Selector DRIFTS accessory and its associated environmental chamber. Spectra of samples (5 wt% in dry KBr) were recorded at 25, 50, 75, 100, 125, 150, 200, 250, 300, 350, 400 and 450 °C. The sample chamber was purged with dry nitrogen whilst 276 scans were collected from 400 to 4000 cm^{-1} at 4 cm^{-1} resolution. Samples were immersed in reagent grade pyridine overnight and the excess pyridine allowed to evapourate before preparing samples and collecting spectra as above. Thermogravimetric traces (7 mg sample weight) were obtained using a Mettler TG50 thermobalance equipped with a TC10A processor.

The fitting of the DRIFTS spectra was accomplished using the Grams software, Galactic Industries Ltd.

3 Results and Discussion

3.1 Kaolin Water Samples

Powder X-ray diffraction traces (not illustrated) confirmed; (i) the presence of a crystalline mica impurity which degraded rapidly with milling time and (ii) that the kaolin became less crystalline as the milling time progressed but that the diagnostic 001 and 002 peaks were still present (at much reduced intensity) after 30 minutes milling.

The thermogravimetric trace for unmilled kaolin is typical in that the layer dehydroxylation occurred as a single event (a loss of 14 wt%) beginning just below 500 °C (Figure 1). The total weight loss increased slightly as the milling time progressed but the weight loss associated with the unchanged kaolin decreased. Moreover, the amount and rate of water loss below 400 °C increased

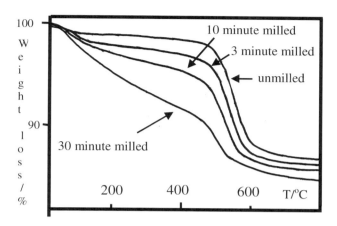

Figure 1 *Thermogravimetric curves for the unrefined and milled kaolins*

with milling time. This behaviour correlates well with the work of others.[1–4]

Figure 2a shows that the unrefined kaolin exhibited four characteristic OH stretching bands associated with a well crystallised kaolin at 3695, 3669, 3650 and 3620 cm^{-1}. As milling time increased, the intensity in the 2800 to 3400 cm^{-1} region, which can be attributed to the presence of H-bonded OH species, increased dramatically. Moreover, the four characteristic bands began to converge and alter in relative intensity after only 3 min milling. After 30 min milling the 3669 and 3650 cm^{-1} bands were no longer visible and the intensity of the hydrogen bonded OH stretching was significantly increased.

The increase in H-bonded OH was reflected in the OH bending region (Figure 2b). The peaks at 1930 and 1823 cm^{-1} are combination bands and can be used as internal markers of intensity. Thus, Figure 2b shows that the intensity in the OH bending region also increased dramatically with milling time. Hence, the DRIFTS spectra recorded at room temperature correlated well with the TG traces shown in Figure 1 in that there was much more adsorbed water present as the milling time progressed. These results are in accord with the results of others.[1–6] Variable temperature DRIFTS revealed, as anticipated, that increasing the sample temperature reduced the intensity in both the OH bending and OH stretching regions, as illustrated for the 30 minute milled sample in Figures 3 and 4. This also emphasised an OH bending band at 1670 cm^{-1} which remained prominent to high temperatures (Figure 3). This band position is indicative of strongly H-bonded water.

This prompted an investigation into whether this band existed in the other milled samples. Consequently, DRIFTS spectra recorded at increasing temperature for each of the samples were fitted and representative fits for the OH stretching (Figure 5 a,b) and OH bending region (Figure 5 c,d) of the 10 min milled sample are presented below. This showed that the 1670 cm^{-1} band (band w in Figure 5 c,d) was present in the 3 and 10 minute milled samples but was more prominent in the latter.

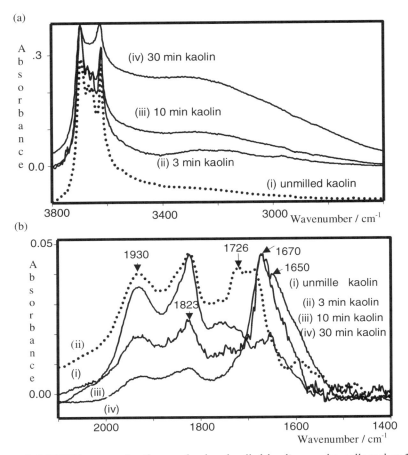

Figure 2 *DRIFTS spectra for the unrefined and milled kaolin samples collected at 25 °C.
(a) OH stretching region; (b) OH bending region*

Figures 5a and 5b illustrate that removal of the broad hydrogen bonded OH stretching band (band g) revealed the presence of an underlying, very strongly hydrogen bonded species centred at 3050 cm^{-1}. The gradual shift in position of the bands in the OH stretching region is illustrated for the 10 min milled sample in this figure along with the associated band intensity (Figures 6). Note that the areas of the bands near 3500 and 2700 cm^{-1} in the 10 minute milled sample do vary with temperature but that the band positions are relatively stable.

The fitting procedure in the OH bending region revealed that the band positions were relatively stable as the temperature increased. Moreover, the band near 1730 cm^{-1} did not diminish in intensity as the temperature was raised which indicates that it too is a structural band as are the 1930 and 1823 cm^{-1} bands. The areas of the OH bending bands for the unmilled and 10 min milled samples are shown in Figure 7.

Gonzalez-Garcia *et al.*,[1] have published electron micrographs which illustrate

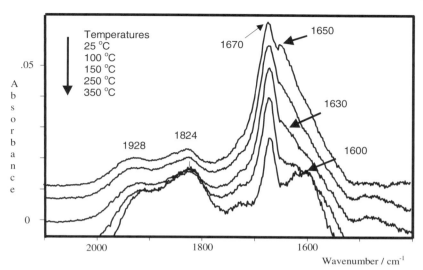

Figure 3 *The effect of temperature on the intensity in the OH bending region for the 30 m milled sample*

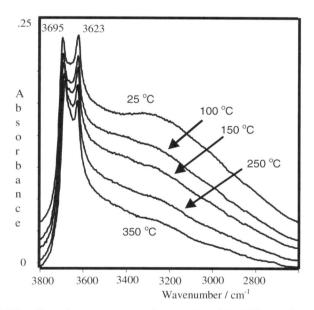

Figure 4 *The effect of temperature on the intensity of the OH stretching region for the 30 m milled sample*

the formation of spherical aggregates of kaolin as the time of milling is increased. These aggregates are made up of kaolin particles which have been reduced in size during the milling process. However, little is known regarding the forces that hold these aggregates together.

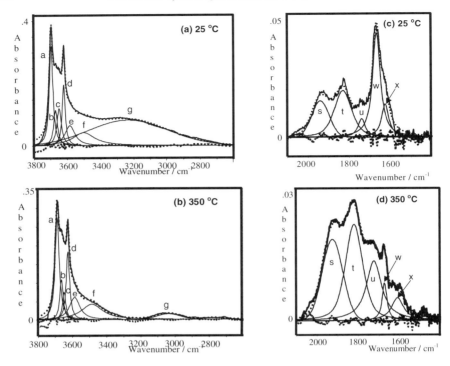

Figure 5 *Representative fitted spectra for the 10 m milled sample. OH stretching region at*
(a) 25 and (b) 350 °C. OH bending region at (c) 25 and (d) 350 °C

The position and thermal stability of the 1670 cm^{-1} band together with the fact that its intensity increases with increased milling time suggests that it may be attributed to water in a very strongly hydrogen bonded environment. It is hypothesised that this band indicates the presence of a type of 'adhesive' water which holds together the aggregates of milled kaolin particles.

3.2 Kaolin Pyridine Samples

The infrared spectrum of pyridine adsorbed to solid surfaces is routinely used for the evaluation of the number and type of acid sites.[7-9] The rich spectrum of adsorbed pyridine can also provide information regarding the extent of hydrogen bonding that the sorbed pyridine is involved in. The presence of Bronsted acid sites is unequivocally identified by the observation of a broad band at 1540 cm^{-1} together with a strong 1490 cm^{-1} band. In contrast the presence of Lewis acid sites are revealed by the presence of a strong, thermally stable 1450 cm^{-1} band which is not to be confused with bands for physisorbed pyridine at 1435 cm^{-1} (easily removed in a flow of nitrogen or a poor vacuum) or H-bonded pyridine at 1440 cm^{-1} (together with a band at 1590 cm^{-1}). The 1490 cm^{-1} band is much weaker for Lewis bound pyridine than for Bronsted bound pyridine.

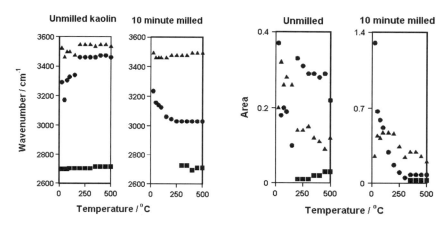

Figure 6 *Comparison of the positions and areas of the H-bonded OH stretching bands for the unmilled and milled samples. The triangles, the circles and the squares correspond to bands f, g and h in the fitted spectra in Figure 4*

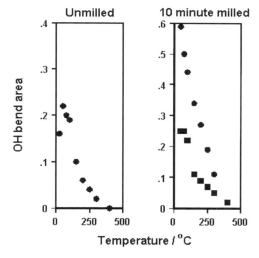

Figure 7 *The areas of the OH bending bands for the unmilled and 10 m milled samples. The filled circles in the unmilled sample represent the area of the 1650 cm^{-1} band, whereas in the 10 m milled sample they represent the area of the 1670 cm^{-1} band. The filled squares indicate the area of the 1630 cm^{-1} band in the 10 m milled sample*

The DRIFTS spectrum for the unmilled kaolin/pyridine sample exhibits bands associated with H-bonded pyridine (1440 and 1593 cm^{-1}) and Bronsted bound pyridine 1540, 1490 and 1609 cm^{-1} (Figure 8). Raising the temperature of the sample resulted in the removal of the bands assigned to H-bonded pyridine and a gradual decrease in intensity of the bands assigned to Bronsted bound pyridine. By 250 °C the spectrum of sorbed pyridine had changed to that of the

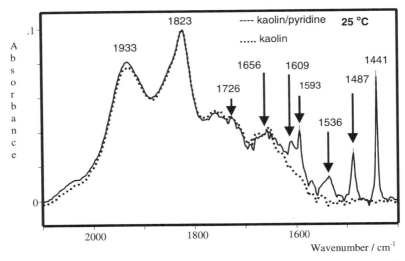

Figure 8 *Comparison of the DRIFTS spectra for unmilled kaolin and unmilled kaolin/pyridine*

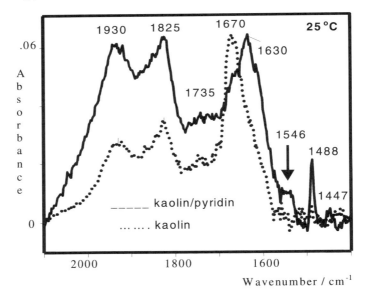

Figure 9 *Comparison of the DRIFTS spectra for 10 m milled kaolin and 10 m milled kaolin/pyridine*

Lewis bound form with a strong band at 1440 cm^{-1} together with a weak 1490 cm^{-1} band. This implies that the Bronsted acidity is associated with the strongly bound water and as this water is removed the pyridine becomes co-ordinated to a Lewis bound site either nearby or at the undercoordinated Al site produced by the removal of surface bound water. This transformation of Bronsted to Lewis acid centres is well established in catalyst chemistry as the sample temperature is raised.[9]

The strong 1490 cm^{-1} band together with the broad peak at 1546 cm^{-1} clearly identifies Bronsted acid sites on the surface of the 10 minute milled sample (Figure 9). In addition the amount of water, particularly the strongly bound ('adhesive') water, which gives rise to the OH bending band at 1670 cm^{-1} was significantly reduced in intensity in the presence of pyridine and the main OH bending band was at 1630 cm^{-1}. This may indicate that the pyridine interacts with the strongly H-bonded water and removes a proton to form the pyridinium ion. As with the unmilled kaolin the intensity of the Bronsted bound pyridine bands decreased as the temperature was increased. At 250 °C there was evidence of strong Lewis bands at 1451 cm^{-1}.

4 Conclusion

Variable temperature DRIFTS analysis of the samples identified the presence of at least two distinct types of sorbed water in the milled samples. In particular, an OH stretching band at 1670 cm^{-1} increased as the milling time increased and exhibited marked thermal stability. The water giving rise to this band has been tentatively attributed to strongly bound water which is holding the ground kaolin particles in larger aggregates which have been observed here[10] and elsewhere.[1] The adsorption of pyridine identified the presence of Bronsted acid sites on the surface of kaolin at room temperature. These were converted to Lewis acid sites as the temperature was raised probably as a result of the thermal desorption of sorbed water.

References

1. F. Gonzalez-Garcia, M.T. Ruiz Abrio and M. Gonzalez Rodriguez, *Clay Miner.*, 1991, **26**, 549.
2. G. Suraj, C.S.P. Iyer, S. Rugmini and Lalithambika, *Appl. Clay Sci.*, 1997, **12**, 111.
3. E. Kristof, A.Z. Juhasz and I. Vassanyi, *Clays Clay Miner.*, 1993, **41**, 608.
4. H. Kodama, L.S. Kotlyar and J.A. Ripmeester, *Clays Clay Miner.*, 1989, **37**, 364.
5. A. La Iglesia and A.J. Aznar, *J. Mater. Sci.*, 1996, **31**, 4671.
6. E. Mendelovici, R. Villalba, A. Sagarzazu and O. Carias, *Clays Clay Miner.*, 1995, **30**, 307.
7. E.P. Parry, *J. Catal.*, 1963, **2**, 317.
8. C. Breen, *Clay Miner.*, 1991, **26**, 473.
9. J.W. Ward, *J. Colloid Interf. Sci.*, 1968, **28**, 269.
10. J. Illés, PhD Thesis, Sheffield Hallam University, 2000.

Computer Simulation of Water Molecules at Mineral Surfaces

M.R. Warne,[1] N.L. Allan[2] and T. Cosgrove[2]

[1]TRIPOS RECEPTOR RESEARCH, BUDE-STRATTON BUSINESS
PARK, BUDE, CORNWALL, EX23 8LY, UK
[2]SCHOOL OF CHEMISTRY, UNIVERSITY OF BRISTOL, CANTOCK'S
CLOSE, CLIFTON, BRISTOL, BS8 1TS, UK

1 Introduction

The nature of the interfacial structure and dynamics between inorganic solids and liquids is of particular interest because of the influence it exerts on the stabilisation properties of industrially important mineral based systems. One of the most common minerals to have been exploited by the paper and ceramics industry is the clay structure of kaolinite. The behaviour of water–kaolinite systems is important since interlayer water acts as a solvent for intercalated species. Henceforth, an understanding of the factors at the atomic level that control the orientation, translation and rotation of water molecules at the mineral surface has implications for processes such as the preparation of pigment dispersions used in paper coatings.

   In these proceedings we summarise the key findings from work which we have published previously.[1] Using 'force field' based simulation approaches we have compared the behaviour of water at both kaolinite and silica surfaces. The effects of changing the ionic strength of the water are also considered. We have examined the spatial motion distribution functions of water molecules away from the mineral surface, along with their diffusion and rotation. Comparison with experimental data is made by linking calculated diffusion coefficients to pulse field gradient nuclear magnetic resonance (PFG NMR) spectroscopy and rotational correlation times to relaxation measurements. NMR is chosen because experimentally it provides a convenient and non-invasive means for measuring molecular motion.

2 Theoretical Methodology

2.1 Static Simulations

To yield the crystal structure and the lattice energy of bulk kaolinite (Figure 1), energy minimisation techniques have been employed. Under three-dimensional periodic boundary conditions, a static energy minimisation of the bulk kaolinite unit cell was performed using the Newton–Raphson method.[2] The calculation was carried out at constant pressure and terminated when the change in energy had been reduced to a level of $9.65 \times 10^{-5}\,\mathrm{kJ\,mol^{-1}}$. The success of any simulation relies on the accuracy and transferability of the short range interatomic potentials. As in a recent study of muscovite,[3] a set of atomic charges and potentials developed by Collins and Catlow[4] have been used which also incorporate the O–H potential added to this set by de Leeuw *et al.*[5] in their study of hydroxylated MgO surfaces.

Kaolinite is easily cleaved perpendicular to the c direction, since the interactions between the aluminosilicate layers are much weaker than the intralayer interactions. Therefore as a part of this work the structure and surface energy, γ_{sc}, of the resulting $\{001\}$ surface are considered. The surface energy may be evaluated from the energy of a single layer surface block of clay in a vacuum, U_s, and the energy of a portion of the bulk clay, U_b, containing the same number of atoms as the surface block.

$$\gamma_{sc} = \frac{U_s - U_b}{A} \qquad (1)$$

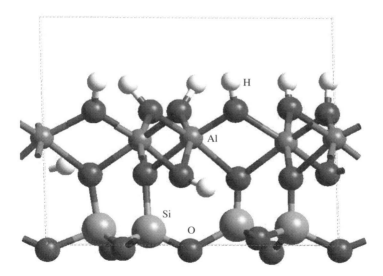

Figure 1 *Structure of kaolinite. The 'Al surface' is the top layer and the 'Si surface' is the bottom layer*

For the atomistic simulation of the kaolinite interface it is assumed that surfaces are planar. Irregularities such as steps, kinks and ledges, which are present on real surfaces, are omitted for the present treatment. For kaolinite the energy of the {001} basal surface was evaluated using a suitable cell containing 425 atoms.

2.2 Molecular Dynamics Simulations

Using the DLPOLY code,[6] molecular dynamics simulations have been performed utilising the NVT ensemble. The surfaces of the minerals considered here have a dipole moment with acts in the direction of the z-axis (type III according to the Tasker classification[7]). To ensure the periodic simulation cell had no net dipole moment, the cells contained either two kaolinite slabs, inverted with respect to each other, consisting of 425 atoms and 820 water molecules (25.81 Å × 26.70 Å × 77.00 Å), or two silica slabs in contact with 800 water molecules (28.50 Å × 28.50 Å × 80.00 Å). The interlayer separation between two kaolinite layers is ≈ 35–40 Å. The same interatomic potentials were used as for the static simulations, allowing all atoms within the clay to move. The charges and potentials for water were those of the TIP3P/OPLS model.[8–11] Interactions between the water molecules and the mineral surface were calculated using the Lorentz–Berthelot geometric mixing rules. For those simulations also including ions in the water layer potentials for Na^+ and Cl^- potentials were extracted from the OPLS forcefield.[11] Short range terms in the interaction potential were cut off at 8.0 Å, while long range coulombic interactions are calculated using the Ewald technique.[12]

Water molecules (density 1 g cm^{-3}) were placed uniformly within the periodic cell. Dynamics simulations were carried out at 298 K for 100 ps. Time steps were set at 0.001 ps and a Berendsen thermostat[13] was used with a time step of 0.5 ps. Typically, 20 ps were used for equilibration runs, followed by production runs of 80 ps.

3 Results

3.1 Static Simulations

Experimental lattice parameters for bulk kaolinite, together with those calculated in the static limit, are listed in Table 1. A difference in the length of parameter b of 2.9% is the largest discrepancy, which is reasonable since our calculation relates to an idealised clay structure.

Calculation of the surface energy, which is an average of the energies of the aluminium and silicon terminated surfaces of kaolinite is 48 mJ m^{-2}. This value is in surprisingly good agreement with values predicted experimentally using adsorption isotherm film pressures[14] (34–103 mJ m^{-2}) and contact angles of solutions on the kaolinite surface[15] (35–96 mJ m^{-2}).

Table 1 *Calculated and experimental lattice parameters for kaolinite*

Lattice parameter	Simulation	Experiment
a (Å)	5.135	5.155
b (Å)	5.305	5.155
c (Å)	7.321	7.405
α (°)	75.05	75.14
β (°)	83.13	84.12
γ (°)	58.87	60.18

3.2 Molecular Dynamics Simulations

3.2.1 Density Profiles. A simulated density profile for water molecules (centre of mass) in one dimension along an axis perpendicular to the clay surface is shown in Figure 2(b). The scale is such that $-5.5 < z < 1$ and $23 < z < 29.5$ corresponds to the clay slabs. Significant structural ordering of the water at both clay surfaces may be observed in Figure 2(b), as is clear by a comparison with Figure 2(a) which shows the density profile for liquid water alone in a periodic box 30 Å in length (density $0.98 \, \mathrm{g \, cm^{-3}}$). The extent of ordering of adsorbed water is in general agreement with experimental data for water at clay surfaces.[16,17] Solid-like structural ordering is observed out to approximately 10 Å from the surface. Figure 2(b) shows that the clay layers are not quite symmetric in the periodic cell. It is interesting to note that in the smaller gap between the clay layers, ordering of the water extends throughout the gap. The appearance of ordering appears to be associated with the relatively ordered arrangement of oxygen atoms and OH groups at the clay surfaces. Figure 3 shows a representative snapshot of the MD simulations carried out for Figure 2(b), which reflects the alignment of the water molecules at the two surfaces. At the Si surface one or both water hydrogens may be involved in H-bonding to the oxygens of the silicate layer. At the Al surface all the atoms in the water molecules can take part in hydrogen bonding to the surface OH groups.

The ordering at the kaolinite surface is striking when compared with similar calculations for the amorphous silica surface (Figure 2c) which shows the ordering is substantially reduced relative to kaolinite. It is tempting to associate this disorder with the disorder at the silica surface; this is consistent with the snapshot in Figure 4.

A change in the structural ordering of the water at the kaolinite and amorphous silica surfaces is observed in Figures 5a and 5b show with increasing ionic strength. In these calculations one Na^+ ion and a charge compensating Cl^- ion were introduced into the solvent. As might be expected the density profile at the kaolinite surface changes markedly, with a significant decrease in the peak intensity close to the surface, but there is still substantial ordering. There are far fewer changes above the amorphous silica surface. Figure 6a shows the density profile for kaolinite including a larger number of ions in the simulation box ($10 \, Na^+$, $10 \, Cl^-$), equivalent to an ionic strength of $\sim 0.4 \, \mathrm{mol \, kg^{-1}}$. The ordering persists even at this much higher concentration, although the number of water

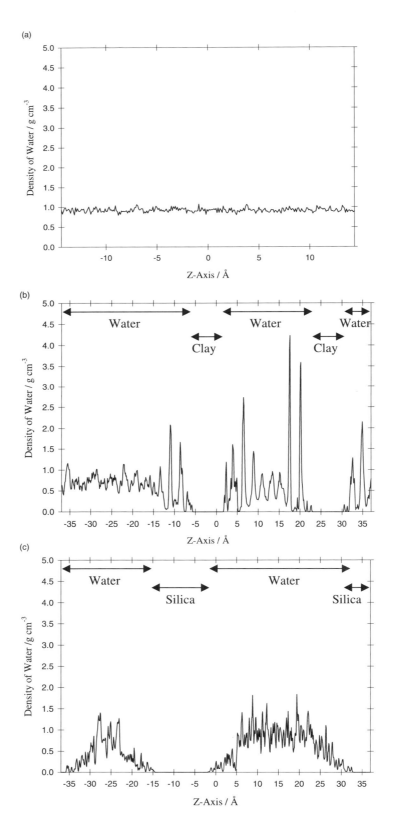

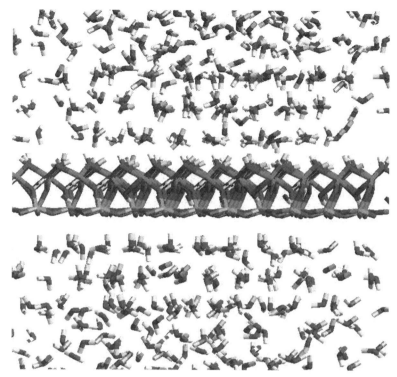

Figure 3 *Representative snapshot of the MD simulations carried out for Figure 2(b)*

molecules at the surface is significantly reduced. The reduced number of molecules at the surface may be accounted for by the displacement of water by Na^+. Figure 6b shows the density profile for amorphous silica with 10 Na^+ and 10 Cl^- ions. Reflecting our previous observations, it can be seen that there is little change in the ordering of the water molecules at the surface.

3.2.2 Self Diffusion Coefficients. Diffusion coefficients may be estimated by exploiting Einstein's relation:

$$D = \frac{\langle |\mathbf{r}_i(t) - \mathbf{r}_i(0)|^2 \rangle}{6t} \tag{2}$$

where the diffusion coefficient of the water molecules, D, is related to the mean square displacement. The value for TIP3P bulk water, $3.9 \times 10^{-9} \, m^2 s^{-1}$, is in

Figure 2 (opposite) (*a*) *One-dimensional density profile of the centres of mass of water molecules in bulk water.* (*b*) *Density profile of the centres of mass of the water molecules along the z-axis (perpendicular to the surface) for kaolinite.* (*c*) *Density profile of the centres of mass of the water molecules along the z-axis (perpendicular to the surface) for amorphous silica*

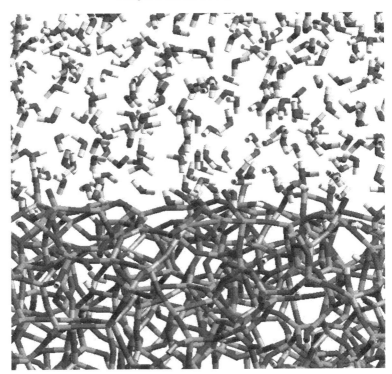

Figure 4 *Representative snapshot of the MD simulations carried out for Figure 2(c)*

agreement with that from previous work $(3.9 \times 10^{-9}\,\mathrm{m^2\,s^{-1}})^2$ but is somewhat in excess of the experimental value of $2.4 \times 10^{-9}\,\mathrm{m^2\,s^{-1}}$.[2] This faster diffusion of simulated water may be a reflection that the density of molecules $(0.98\,\mathrm{g\,cm^{-3}})$ in the periodic cell is slightly lower than that found experimentally $(1.00\,\mathrm{g\,cm^{-3}})$.

For both kaolinite and amorphous silica surfaces, the average diffusion coefficients for all water molecules in the simulation cell drop by over an order of magnitude, reflecting the marked ordering of the water molecules. Assuming that fast exchange of water between the bulk and surface regions occurs,[18] it would be expected that diffusion of water molecules in the surface region was even slower than the average value for the simulation box. Increasing the ionic strength of the solution has a much smaller effect on the average diffusion coefficients, for both kaolinite and silica, as shown by the remaining entries in Table 2.

3.2.3 Dynamics of Reorientation. The approach of Impey, Madden and McDonald[19] is used to consider the dynamics of reorientation for water molecules. The time autocorrelation function of the second Legrendre polynomial P_2 of the angle subtended by the intramolecular H–H bond vector at time t with respect to its position at time $t = 0$ is calculated:

$$C_2^{H-H}(t) = \langle P_2[\mathbf{e}^{H-H}(t)\cdot \mathbf{e}^{H-H}(0)]\rangle \tag{3}$$

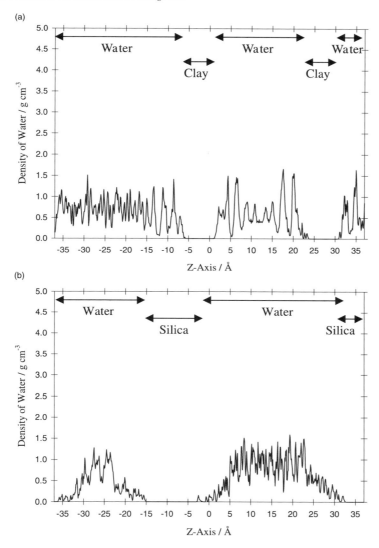

Figure 5 (a) *Density profile of the centres of mass of the water molecules along the z-axis (perpendicular to the surface) for kaolinite with one* Na^+ *and one* Cl^- *added to the simulation box.* (b) *Density profile of the centres of mass of the water molecules along the z-axis (perpendicular to the surface) for amorphous silica with one* Na^+ *and one* Cl^- *added to the simulation box*

e is a unit vector with the same direction as the H–H interatomic vector and $\langle \rangle$ represents the average over all molecules at time t. At large t (> 5 ps) least squares fitting of $C_2^{H-H}(t)$ to the exponential function $A^p \exp(-t/t_C)$ allows us to determine the rotational correlation time, τ_c, given by $A_p t_c$. A value for τ_c of 2.2 ps is predicted for TIP3P water, slightly lower than the value of 2.7 ps predicted by experiment.[20]

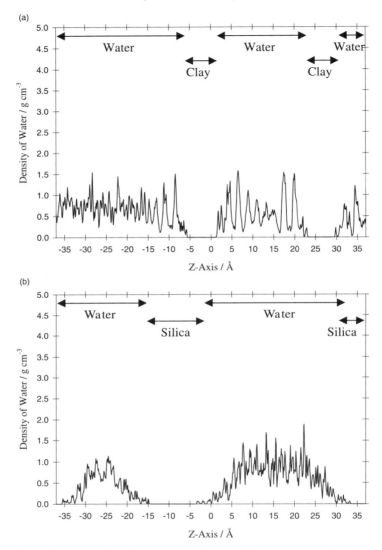

Figure 6 (a) *Density profile of the centres of mass of the water molecules along the z-axis*
(perpendicular to the surface) for kaolinite with ten Na^+ and ten Cl^- added to the
simulation box. (b) Density profile of the centres of mass of the water molecules
along the z-axis (perpendicular to the surface) for amorphous silica with ten Na^+
and ten Cl^- added to the simulation box

Rotational correlation times are listed in Table 3. As with the diffusion
coefficients, the effect of the ordering imposed by the clay surface is clearly
evident.

3.2.4 Average NMR Relaxation Times.   The NMR relaxation time for water is
predicted using the approach of Carrington and McLachlan.[20] Pure water can be

Table 2 *Average self diffusion coefficients* (D) *of water in various environments*

Model	Simulated Diffusion Coefficient ($m^2 s^{-1}$)
TIP3P Bulk Water	3.9×10^{-9}
Water at the kaolinite interface	9.6×10^{-11}
Water at the kaolinite interface + 1 (Na^+Cl^-)	7.5×10^{-11}
Water at the kaolinite interface + 10 (Na^+Cl^-)	6.9×10^{-11}
Water at the silica surface	2.6×10^{-11}
Water at the silica surface + 1 (Na^+Cl^-)	2.7×10^{-11}
Water at the silica surface + 10 (Na^+Cl^-)	3.1×10^{-11}

Table 3 *Average rotational correlation functions* (τ^c) *of water in various environments*

Model	Rotational correlation function (ps)
TIP3P Bulk Water	2.2
Water at the kaolinite interface	23.3
Water at the kaolinite interface + 1 (Na^+Cl^-)	33.5
Water at the kaolinite interface + 10 (Na^+Cl^-)	31.9
Water at the silica surface	85.1
Water at the silica surface + 1 (Na^+Cl^-)	95.5
Water at the silica surface + 10 (Na^+Cl^-)	81.0

described quantum mechanically as a particle with $I = 1$ since the total spin angular momentum is conserved during molecular tumbling. For $\omega^2\tau_c^2 \ll 1$, (where ω is the angular velocity and τ_c is the correlation time) the relaxation time has two contributions, one from the rotation of the molecule (T_{1rot}) and one from the translation (T_{1trans}). Using perturbation theory, the relaxation or longitudinal magnetisation time, T_1, of the water molecule is then given by:

$$\frac{1}{T_1} = \frac{1}{T_{1rot}} + \frac{1}{T_{1trans}} \tag{4}$$

The rotation of the molecule causes a magnetic dipole–dipole interaction between the protons in the molecule, which is a function of the distance r between the two dipoles. The relaxation rate, ($1/T_{1rot}$), varies linearly with the rotation correlation coefficient, τ_c.

$$\frac{1}{T_{1rot}} = \left(\frac{\mu_0}{4\pi}\right)^2 \frac{3}{2} \frac{\gamma_{1H}^4 \hbar^2 \tau_c}{r^6} \tag{5}$$

where μ_0 is the permeability constant and γ_{1H} is the gyromagnetic constant.

Water molecules diffusing past each other cause further intermolecular magnetic dipole–dipole interactions. These are a function of the concentration of spins, N, the average diffusion coefficient, D, and the distance of closest approach between the spins, b.

Table 4 *Estimated average NMR relaxation times of water in various environments*

Model	NMR Relaxation time (s)
TIP3P Bulk Water	6.39
Water at the kaolinite interface	0.37
Water at the kaolinite interface + 1 (Na^+Cl^-)	0.27
Water at the kaolinite interface + 10 (Na^+Cl^-)	0.26
Water at the silica surface	0.10
Water at the silica surface + 1 (Na^+Cl^-)	0.10
Water at the silica surface + 10 (Na^+Cl^-)	0.11

$$\frac{1}{T_{1\text{trans}}} = \left(\frac{\mu_0}{4\pi}\right)^2 \frac{\pi}{5} \frac{\gamma^4 h^2 N}{Db} \tag{6}$$

By taking the simulated values for D and τ_c, as well as $N = 6.75 \times 10^{28}\,\text{m}^{-3}$ and $b = 1.74\,\text{Å}$, consistent with the size of the simulation box and the number of water molecules, the average NMR relaxation time can be estimated. T_1 is predicted to be 6.39 s, an overestimate of the experimental value of 3.3 s.[20] Although the value is overestimated, in the long term the interest in T_1 is going to be chiefly in how it varies from system to system, giving qualitative values as opposed to quantitative ones.

Values for T_1 (Table 4) are an average of both water molecules at the surface and water molecules in the bulk.

4 Discussion and Conclusions

Good agreement between the simulated lattice parameters for kaolinite and those observed experimentally along with a fair reproduction of the dynamic properties of water, means we can be confident in our model for investigating NMR relaxation times. From the simulations a substantial decrease in the average diffusion coefficient, an increase in correlation time and hence a decrease in relaxation time of the water is seen upon the introduction of both kaolinite and silica surfaces, in agreement with experimental observations. There would appear to be a correlation between the change in magnitude of the diffusion coefficient and the change in magnitude of the rotational correlation time for each system in the 2-D region, implying that the motions are not completely independent.

The marked difference in the relaxation times for the kaolinite and silica may be attributed to the nature of the surface. Intuitively, the hydrogen bonding which influences the increased structure at the kaolinite surface would be expected to give shorter values for the relaxation time. However this is not observed in the simulations. Instead, shorter values are seen for the silica surface which is a result of water molecules becoming trapped in the cage-like amorphous silica surface. This reflects experimental results where precipitated silica surfaces are microporous and water inclusion in the surface is common.

The addition of ions to both the kaolinite and silica systems show different trends. Increasing the ionic strength in the kaolinite system results in water molecules being displaced from the surface. The long range order of the water in the bulk is disturbed, as seen in the density profiles, however a high degree of structure remains at the surface, the driving forces of which are the kaolinite itself and the hydration shells around the ions. The silica surfaces appear less effected by ionic strength, which is reflected in both the values of T_1 and the density profiles. This is attributed to the lower fraction of OH groups at the surface, reducing the tendency for ions to adsorb and create local ordering.

The ratio of the values for the NMR relaxation times in water[20] and at the silica surface are in good agreement with that observed experimentally.[21] Effects due to ionic strength are minor. For kaolinite there is no published experimental data for comparison due to problems with uneven coagulation. However, predicted values for kaolinite are of the same order as for silica.

References

1. M.R. Warne, N.L. Allan and T. Cosgrove, *Phys. Chem. Chem. Phys.*, 2000, **2**, 3663.
2. W.H. Press, S.A. Teukolsky, W.T. Vetterling and B.P. Flannery, *Numerical Recipes in FORTRAN. The Art of Scientific Computing*, 2nd ed., Cambridge University Press, 1992.
3. J.A. Purton, N.L. Allan and J.D. Blundy, *J. Mater. Chem.*, 1997, **7**, 1947.
4. D.R. Collins and C.R.A. Catlow, *Am. Mineral.*, 1992, **77**, 1172.
5. N.H. de Leeuw, G.W. Watson and S.C. Parker, *J. Chem. Soc., Faraday Trans.*, 1996, **92**, 2081.
6. T.R. Forester and W. Smith, CCLRC, Daresbury Laboratory, Warrington, England, 1995.
7. P.W. Tasker, *Philos. Mag.*, 1979, **39**, 119.
8. T.R. Forester and W. Smith, CCLRC, Daresbury Laboratory, Warrington, England, 1995.
9. W.L. Jorgensen, J. Chandrasekhar, J.D. Madura, R.W. Impey and M.L. Klein, *J. Chem. Phys.*, 1983, **79**, 926.
10. S.J. Weiner, P.A. Kollman, D.T. Nguyen and D.A. Case, *J. Comp. Chem.*, 1986, **7**, 230.
11. W.L. Jorgensen, D.S. Maxwell and J. Tirado-Rives, *J. Am. Chem. Soc.*, 1996, **118**, 11 225.
12. J.M. Ziman, *Principles of the Theory of Solids*, Cambridge University Press, Cambridge, 1964.
13. H.J.C. Berendsen, J.P.M. Postma, W.F. van Gunsteren, A. di Nola and J.R. Haak, *J. Chem. Phys.*, 1984, **81**, 3684.
14. E. Chibowski and P. Staszczuk, *Clays Clay Miner.*, 1988, **36**, 455.
15. B. Janczuk, E.M.H. Chibowski, T. Bialopiotrowicz and J. Stawinski, *Clays Clay Miner.*, 1989, **37**, 269.
16. R.E. Grim, *Clay Mineralogy*, 2nd ed., McGraw-Hill, New York, 1968.
17. A. Delville and M. Letellier, *Langmuir*, 1995, **11**, 1361.
18. G.P van der Beek, M.A. Cohen Stuart and T. Cosgrove, *Langmuir*, 1991, **7**, 327.
19. R.W. Impey, P.A. Madden and I.R. McDonald, *Mol. Phys.*, 1982, **46**, 513.
20. A. Carrington and A.D. McLachlan, *Introduction to Magnetic Resonance*, Chapman and Hall, New York, 1980.
21. H. Pfeifer, *NMR: Basic Princ. and Prog.*, 1972, **7**, 55.

Modified Hydroxamate Collectors for Kaolin Flotation

C. Basilio,[1] R.A. Lowe,[1] A. Gorken,[2] L. Magliocco[2] and R. Hagy[3]

[1]THIELE KAOLIN COMPANY, SANDERSVILLE, GEORGIA, USA
[2]CYTEC INDUSTRIES INC., STAMFORD, CONNECTICUT, USA
[3]CYTEC INDUSTRIES INC., SOUTH CHARLESTON, WEST VIRGINIA, USA

1 Introduction

Kaolin clays are naturally occuring sedimentary deposits composed largely of kaolinite mineral. Typical impurities in these deposits are iron oxides, titaniferous minerals, silica, feldspar, mica, sulfides and organic matter. The majority of kaolin clay produced in the world is used in the paper industry as coating and filler materials. This mineral also makes an excellent filler, carrier, opacifier and diluent in a variety of industrial products such as paints, plastics, cement, rubber, pharmaceuticals, *etc*.

To make kaolin acceptable for use, crude kaolin deposits require beneficiation to remove the associated impurities. The colored impurities, specifically titaniferous minerals and iron oxides, are generally removed by reverse froth flotation, high gradient magnetic separation, selective flocculation and/or leaching.

Froth flotation has proven to be an efficient method of removing titaniferous impurities (mainly iron-rich anatase) from kaolin clays. Fatty acid reagent, primarily tall oil, is used extensively in the reverse flotation of these impurities. This flotation collector typically requires divalent cations (usually Ca^{2+}) to activate the coloured impurities and enhance collector adsorption. This is not very selective since the tall oil can also absorb on the kaolinite particles. Alkyl hydroxamate collectors are relatively new in the kaolin industry but provide significant advantages.[1] Hydroxamates do not require activators, substantially increase the removal of colored impurities and are very selective.

In this study, the flotation performance of a series of newly developed hydroxamate-based collectors was evaluated on different crude clays from Georgia, USA. These new collectors provide improved selectivity over the standard tall oil chemistry and the commercial hydroxamate reagent, Aero Promoter 6493 (AP 6493). The modified hydroxamates have the advantages of higher activity, easier handling, greater stability, increased frothability if desired and can be tailored *via* the use of different carriers and methods of synthesising the reagent.

2 Experimental

2.1 Reagents

The newly developed modified alkyl hydroxamate reagents (from Cytec Industries) tested in this study include S-8704, S-8704D, S-8705, S-8706, S-8706D and S-8765, while the reference collectors used were Aero Promoter 6493 (also from Cytec) and tall oil. The crude clays were dispersed using sodium silicate while soda ash was used to adjust the pH. The frother used in the flotation tests was Aerofroth 70.

2.2 Clay sample

Tests were conducted on 'soft' and 'hard' kaolin clays from Georgia, USA. Soft kaolins are relatively coarse-grained (coarser than 65% < 2 microns), found in Cretaceous age strata, and typically have low TiO_2 contents (1–3%). The hard kaolins are fine-grained (finer than 80% < 2 microns), found in Eocene age strata and have higher TiO_2 contents (2–8%).

2.3 Procedure

The laboratory test procedure used in the different flotation tests is basically composed of three stages: blunging, conditioning and flotation. Blunging (or dispersion) of the crude clay is carried out in a Cowles Dissolver at 60% solids (by weight) using sodium silicate as the dispersant and soda ash (to pH 9) as the pH modifier. After the clay has been dispersed, conditioning is conducted in the same vessel using the required amount of collector. After the conditioning step, the pulp temperature and pH are recorded and the pulp is diluted to 25% (by weight) with water and transferred to the flotation cell. The required amount of frother is then added to the slurry and mixed for a minute prior to flotation. Flotation tests are carried out either in a batch Denver cell or laboratory-sized flotation column (7.62 cm diameter). The froth reject and the slurry remaining in the cell (or underflow in the flotation column) were collected, dried, weighed and analysed for TiO_2 using XRF spectroscopy. Plant trial tests were conducted at Thiele Kaolin Company's flotation plant employing Microcel flotation columns.

The flotation responses of the kaolin samples to the various flotation collectors were measured using the separation efficiency (SE) index. This index combines both grade and recovery to describe the efficiency of the beneficiation process.[2] The mathematical expression used to compute the separation efficiency is the following:

$$SE = Rm - Rg \qquad (1)$$

where Rm is the percent recovery of the valuable material (kaolin) and Rg is the percent recovery of the gangue (TiO_2) into the product.

3 Results and Discussion

Reagent screening tests were conducted on the different modified hydroxamate reagents. The alkyl hydroxamate activities of the modified reagents are higher than that of AP 6493 (*i.e.* > 30% active). These reagents also have different carrier solvents that allow the alkyl hydroxamate to remain soluble at ambient temperatures. In addition, some of the reagents use a carrier solvent that has some frothing properties; thus reducing the frother requirement for flotation.

Flotation tests were conducted on a soft kaolin sample that is known to have good response to flotation (sample A). A laboratory Denver cell was used in these tests and the results are presented in Table 1. Using tall oil as a collector, the flotation performance is relatively poor, as shown by the low flotation recovery. This shows the poor selectivity of tall oil in kaolin flotation. Another test was conducted using the flotation chemistry given in US Patent 5522986.[3] This invention uses a blend of alkyl hydroxamate and tall oil as the collector. The results for this chemistry (AP 6493/tall oil blend) show good flotation recovery and a significant reduction in TiO_2. The use of a higher AP 6493 dosage without tall oil added produced a significantly lower TiO_2 grade at comparable flotation recovery, thereby increasing the SE index.

For the modified reagents, the use of S-8765 resulted in poorer flotation performance. A relatively high S-8765 dosage is required to obtain comparable results. However, S-8704, S-8705 and S-8706 gave significantly improved TiO_2 rejections at equal to higher clay recoveries. Reagents S-8704 and S-8706 showed the most promise of the new collectors. S-8706 has a higher hydroxamate activity than AP 6493 and uses a carrier solvent with some frothing properties. This collector produced a considerably lower TiO_2 grade with very high recovery of 96%. The separation efficiency for S-8706 is almost twice that for the AP 6493/tall oil blend.

Table 1 *Denver cell flotation testing of sample A (soft kaolin with good flotation response)*

Collector	Dosage (kg ton^{-1})	Product % TiO_2	Clay % recovery	Separation efficiency
AP 6493/Tall oil	0.5/0.5	0.84	84	40.5
AP 6493	1	0.27	80	66
Tall oil	1	0.79	67	32
S-8765	1	0.80	81	41
S-8765	2	0.54	84	56
S-8704	0.5	0.55	96	64
S-8704	1	0.44	86	62
S-8705	0.75	0.50	84	58
S-8706	0.625	0.33	96	75

Table 2 *Denver cell flotation testing of sample B (soft kaolin with poor flotation response)*

Collector	Dosage (kg ton^{-1})	Product % TiO_2	Clay % recovery	Separation efficiency
AP 6493	0.5	0.25	47	41
Tall oil	2	0.48	38	28
S-8704	0.5	0.26	47	40
S-8704D	0.5	0.24	53	47
S-8706D	0.5	0.39	60	47

Table 3 *Denver cell flotation testing of sample C (soft kaolin with high TiO_2 and Fe content)*

Collector	Product % TiO_2	Clay % recovery	Separation efficiency
AP 6493	0.81	71	39
Tall oil	1.06	60	25
S-8704	0.53	58	41
S-8706D	0.66	64	41

Table 4 *Denver cell flotation testing of sample D (hard kaolin) using S-8704 and S-8706*

Collector	Product % TiO_2	Clay % recovery	Separation efficiency
AP 6493	1.37	61	28
Tall oil	1.73	68	20
S-8704	1.17	45	24
S-8706	0.96	61	38

The next series of tests was carried out on a soft kaolin sample that is known to respond poorly to flotation (sample B). Tests were carried out using S-8704 and S-8706, the two most promising reagents found for sample A. Table 2 shows that a very low product TiO_2 grade was obtained with AP 6493 but the clay yield was poor. The flotation performance was even poorer for the sample floated with tall oil. The modified hydroxamate, S-8704, gave a flotation performance identical to that for AP 6493. On the other hand, reagents S-8704D and S-8706D gave significantly higher clay recoveries with almost similar amounts of TiO^2 removal; thus, giving higher SE values.

Tests were also conducted on a soft kaolin sample with relatively higher TiO_2 and iron content (sample C) than the other soft kaolin samples. Table 3 shows that flotation with tall oil gives poor flotation performance. Using AP 6493, the grade and recovery are both improved resulting in a higher separation efficiency. The modified hydroxamates, S-8704 and S-8706D, give significantly higher TiO_2 removal but with slightly lower clay yields. Higher collector dosages are prob-

Table 5 *Column flotation testing of sample A using S-8706D and S-8706*

Collector	Dosage (kg ton^{-1})	Separation efficiency
AP 6493	1	66
AP 6493/Tall oil	0.5/0.5	67
S-8706	0.625	73
S-8706/Tall oil	0.5/0.5	70
S-8706D	0.75	66

Table 6 *Plant trial data – soft kaolin using S-8706*

Collector	Average separation efficiency
AP 6493 (Before trial)	70
S-8706	73
AP 6493 (After trial)	68

ably required to improve the flotation response of sample C.

Hard kaolins from East Georgia, USA are known to give poor response to flotation.[4] The efficacy of S-8704 and S-8706 are tested on sample D, a fine, hard kaolin with TiO_2 content of 2.45% and particle size distribution of about 88% < 2 microns. Using tall oil as the collector, the product grade is still high and the resulting separation efficiency is low (see Table 4). Results for AP 6493 show improved anatase removal and a resulting increase in SE value. Reagent S-8706 gives further improvement in anatase removal but not S-8704. However, the SE values are all relatively low, showing the poor flotation response of fine-grained kaolin clays.

The results for the different crude samples show that S-8706 and S-8706D are the most promising of the different collectors tested. These reagents showed improved TiO_2 removal with relatively good clay recovery for either soft or hard kaolins. To further test the efficacy of these reagents, column flotation tests were conducted on Sample A. The results of the different flotation tests are presented in Table 5. As shown, significantly higher SE values were obtained with the use of the flotation column (see Table 1). Using S-8706 alone, improved flotation performance is obtained at a lower collector dosage. In the case of the hydroxamate/tall oil blend, replacing AP 6493 with S-8706 results in an increase in SE value. Flotation with S-8706D, a derivative of S-8706 using a different carrier solvent, results in similar flotation performance as AP 6493 but requires a lower collector dosage. These results show the improved flotation performance of these modified hydroxamates.

To further prove the efficacy of S-8706 and its applicability on a full-scale, a plant trial was conducted at Thiele Kaolin Company's flotation plant. The average SE value obtained with AP 6493 before the trial run was 70. Replacing AP 6493 with S-8706, the flotation performance during the trial run improved to

73. After the plant trial run with S-8706, the measured SE value for AP 6493 averaged about 68. This trial shows the improved flotation performance obtained with the use of S-8706 (Table 6).

4 Conclusions

New modified hydroxamate collectors were developed with the following improved properties: higher activity, easier handling, improved stability and can have increased frothing property. Flotation tests were conducted on different soft and hard kaolin crude samples from Georgia, USA. Laboratory screening tests were conducted to identify the most promising modified hydroxamate reagent. The results show that S-8704 and S-8706 give better flotation performances than AP 6493 for coarse-grained soft and fine-grained hard Georgia kaolins. In a full-scale plant trial using flotation columns, S-8706 showed better flotation performance in the flotation of soft Georgia kaolins than the current commercially available reagent, AP 6493. Reagent S-8706 has a higher hydroxamate activity and better stability (*i.e.* remained liquid at ambient temperature) than AP 6493.

References

1. R.H. Yoon and J. Yordan, 'Beneficiation of Kaolin Clay by Froth Flotation Using Hydroxamate Collectors', *Minerals Engineering*, 1992, **5**(3–5), pp. 457–467.
2. B. Wills, in *Mineral Processing Technology, Fifth edition*, Pergamon Press, Tarrytown, New York, 1992, pp. 33–35.
3. J. Shi and J. Yordan, '*Process for Removing Impurities from Kaolin Clays*', US Pat. 5522986, 1996.
4. J. Yordan, R.H. Yoon and T. Hilderbrand, 'Hydroxamate *vs.* Fatty Acid Flotation for the Beneficiation of Georgia Kaolin' in *Reagents to Better Metallurgy*, Society for Mining Engineering, 1994, Chapter 22.

Microbiological Control

The Biocidal Products Directive and the Mineral Processing Industry: Possible Issues and Implications

J. Duddridge

ROHM AND HAAS FRANCE SAS, EUROPEAN LABORATORIES, SOPHIA ANTIPOLIS, 06560 VALBONNE, FRANCE

1 Summary

Directive 98/8/EC, the so-called Biocidal Products Directive (BPD) was integrated into each Member State National legislation for May 14th 2000. Within the Annexes of this directive several biocidal product types are identified that are relevant either directly or indirectly to mineral processing or to the use of mineral slurries. This legislation will have an increasingly significant impact on the active substance producers, the biocidal product formulators and the end-user industries they serve. Some of the more important consequences/issues for the industry include:

- The gradual phase out of non-notified or unsupported active substances.
- The general industry charges to cover the authorisation/registration costs of the Competent Authorities could lead to extensive product line rationalisation in many formulator companies.
- A likely outcome of these last two points is that there will be less choice (fewer actives/fewer formulations) and ultimately higher costs for the consumer. In certain cases there may be a higher risk of functional protection failure and hygiene decline in some product types.
- Many end-user processes and products may need to be modified or reformulated to accept the active substances and formulations remaining on the market.
- The Technical Guidance Documents supporting the directive are confusing, overlapping, inflexible, incomplete and frequently lack pragmatic guidance. Furthermore, their legal status is also not well established in some Member States. The required level of Community harmonisation may not be easy to achieve under these conditions.
- The data requirements are in excess of those required for an acceptable risk

assessment and are so extensive/costly that they will exclude market access to many SMEs.

- Appropriate tools essential to the development of the supporting dossiers are missing (*e.g.* realistic human and environmental exposure scenarios, agreed emission models, leach tests, environmental fate methods, efficacy methods and environmental monitoring guidelines) for many of the biocidal product types. Without these practical tools the risk assessment process and therefore the directive, will be unworkable for industry in certain product types. Industry will be obliged to work with the authorities to develop methods to cover these technical gaps. Given the international nature of many of these issues (*e.g.* many are already part of OECD pesticide initiatives) resource needs to cover this work will be extensive and may take several years to arrive at agreed international standards.
- This burden of high cost against such a background of confusion will stifle innovation. Very few new active substances will be brought to the market in the coming years. This applies equally to alternative methods of microbial control that fall within the directives definition of a biocidal product *i.e.* 'intended to destroy, deter, render harmless, prevent the action of, or otherwise exert a controlling effect on any harmful organism by chemical or biological means'.

Many of these issues can only be resolved by a high level of collaboration between active substance producers, the biocidal product formulators, the service companies and the final customer.

2 Biocides and the Mineral Processing Industries

The susceptibility of unprotected aqueous-based mineral slurry processes, suspension aids and the slurries themselves to bacterial and fungal contamination, growth and ultimate spoilage is a well documented phenomenon. Many of the organic and inorganic components of slurry formulations fulfill the nutritional requirements of the microorganisms while machine and process operating characteristics (temperature, pH, mixing, aeration, wetted surfaces) provide fully acceptable growth conditions. Excessive microbial growth in these slurries affects either directly or indirectly their functional performance and under certain circumstances the hygienic operation of the plant.

Biocides are essential raw materials in the manufacture and formulation of mineral slurries. It is well established that they must be added to prevent the inevitable growth of spoilage organisms in the processing of the mineral slurries and their storage at the site of production or use. Biocides also avoid contamination with potentially harmful pathogenic species that could represent a health risk. It is only in this way that the functional, economical and safe performance of the slurries can be assured. Biocides (mostly fungicides) may also be added to mineral slurry formulations to impart anti-fungal protection to end-use coated products *e.g.* paper used for wine labels.

3 The Biocidal Products Directive . . . Some Background

Directive 98/8/EC, the so-called Biocidal Products Directive (BPD) was published in 1998 and was required to have been integrated into each Member State national legislation by May 14th 2000. The BPD harmonises the regulation of active substances at a European level through the development of a positive list of active substances (*i.e.* Annexes 1, 1A and 1B) that can be used in biocidal products and the authorisation/registration of biocidal products derived from those active substances throughout the Member States. It is proposed that harmonisation is achieved through common guidance to:

1. Producers/notifiers and Competent Authorities defining the data and studies required for the inclusion of an active substance onto one of the positive lists (Annexes IIA and IIIA).
2. Competent Authorities defining criteria for the inclusion of active substances onto one of the positive lists.
3. Producers/formulators and Competent Authorities defining the data and studies required for authorisation of a biocidal product (Annexes IIB and IIIB).
4. Competent Authorities on how to perform the administrative and scientific evaluation of applications for the authorisation/registration of biocidal products (the 'common principles', Annex VI).

Within this context three Technical Notes for Guidance (TNGs) were commissioned for preparation by May 14th 2000. Only one document is currently available, the *'Guidance on Data Requirements for Active Substances and Biocidal Products'*. This was finalised in December 1999 and is available on the ECB website. The two other TNGs (namely, the Criteria for Annex 1 Entry and the Common Principles for the Evaluation of Dossiers) appeared as draft documents in 1998. Revised versions of these documents have since been made available for comment. No indication has been given by the Commission when final versions of the documents will be published.

4 The BPD and the Mineral Processing Industries

Annex V of the directive specifies 23 product types. Of these, product type 06 (in-can preservatives) and product type 12 (slimicides) are probably the most relevant to the mineral processing industries and the users of their products. Under certain circumstances product type 07 (film preservatives) and product type 09 (fibre, leather, rubber and polymerised material preservatives) may also be relevant.

This carries with it specific obligations for active substance producers and biocidal product formulators with respect to:

• The data and studies needed for any active substance to be notified or biocidal product to be authorised in this product type.
• The demonstration of acceptable risk for humans, animals and the environ-

ment resulting from the application, use and disposal of biocidal products or treated articles in this product type.
• The demonstration that the biocidal product is sufficiently efficacious within the context of the proposed label claim and directions of use.

5 The Implications of the BPD 1st Review Regulation (Identification/Notification Process)

Issue: The major implication of this 1st review regulation is that between mid–end 2002 to 2006 there will be a gradual removal/phase out of non-identified, non-notified active substances and the biocidal products developed from them. It is impossible to say how many actives/biocidal products will be removed from the market by 2006. It is clear however, that a significant level of biocidal product reformulation may have to take place over the coming years.

The intention of the Commissions 1st Review Regulation is to lay out the ground rules and data/information requirements associated with the identification and notification of existing active substances within the BPD. It has prioritised the first set of active substances (in this case Wood Preservatives and Rodenticides) to be assessed for entry onto Annex 1. The regulation is aimed at collecting the necessary input from producers and formulators to:

• Develop a list of (1) identified existing active substances, (2) notified existing active substances. These lists will be published at a later date by the Commission.
• Develop a prioritisation base for assessment call-in of notified substances. This will be part of the 2nd review regulation.

The 1st review regulation (EC No. 1896/2000) was published on 8th September 2000 and came into force on 28th September 2000.

Industry then has 18 months to identify/notify and 42 months to submit full dossiers for wood preservatives or rodenticides.

Implications of identification/notification include:

• Non-identified active substances cannot be sold after the publication of the list of identified/notified existing substances (2nd Review Regulation, April/May 2003?).
• Identified but non-notified substances will be phased out within 3 years (?) of the publication of the list of identified/notified existing substances.
• Biocidal products in non-notified product types (of notified active substances) will be phased out within 3 years (?) of the publication of the list of identified/notified existing substances.
• Notified active substances can be sold into notified product types up to and after entry onto Annex 1. If an active substance is not accepted onto Annex 1 it will then be phased out (say 3 years after the assessment ruling).

The major implication of this 1st review regulation is that between 2003 to 2006 there will be a gradual removal/phase out of non-identified, non-notified active substances and the biocidal products developed from them. Thereafter, active substances refused Annex 1 entry will be progressively removed from the market. Industry charges levied for the assessment of the dossiers in all Member States will also lead to extensive product line rationalisation (*i.e.* fewer biocidal products). It is difficult to say how many active substances/biocidal products will be removed from the market by 2006. However, CEFIC has estimated that roughly 60% of the *ca.* 2000 existing active substances and the *ca.* 10–20 000 biocidal products marketed in the EU will cease to be commercialised as a result of the directive. Undoubtedly the industries supported by these biocides may be presented with the following potential issues; (1) a decrease in functional protection, (2) hygiene decline, resistance/tolerance, (3) process modifications, (4) end-product re-formulation and (5) higher costs.

The review regulation also requires that co-notifiers for the same active substance *'shall undertake all reasonable efforts to present a common notification, in whole or in part, in order to minimise animal testing'* for Annex 1 assessment. Unfortunately, this request for industrial collaboration must also be considered within the context that the *'specific provisions'* proposed to be developed by the Commission are not yet known and that the current status of data protection and confidentiality within the directive does not really encourage either investment or collaboration.

6 Implications of the Technical Notes for Guidance

> *Issue:* In their current state these TNGs are complex, often overlapping, difficult to use at a practical level and often contradictory (*e.g.* the KEMI document bases certain decision criteria for entry onto Annex 1 on hazard rather than risk).

The data requirements outlined in the *'Guidance on Data Requirements for Active Substances and Biocidal Products'* are extensive, often going beyond what is reasonably needed to do an adequate risk assessment. The complexity and cost of the data requirements will undoubtedly exclude market access to many SMEs. A recent opinion document[5] from the Scientific Committee on Toxicity, Ecotoxicity and the Environment (CSTEE) of DG24 suggests that it would have been preferable to have a smaller core set for the identification of human health hazards and that further studies should be justified on the basis of anticipated human exposure and the results from the core set. Similarly the data requirements for the assessment of environmental hazards should be aligned to expected exposure scenarios arising from the 23 product types. In other words, the data requirements should have been exposure driven. Unfortunately, predictive operator exposure models and environmental emission scenarios do not exist for many product types. This gap is recognised by the Commission and work is underway to develop the appropriate human and environmental exposure scenarios. This TNG is also incomplete and frequently lacks pragmatic guidance. Appropriate tools essential to the development of the supporting dossiers are missing (*e.g.* realistic human and environmental exposure scenarios, agreed emission models, leach tests, environmental fate methods, efficacy methods and environmental monitoring guidelines) for many of the biocidal product types. Without these practical tools, the risk assessment process and therefore the directive, will be unworkable for industry in certain product types. Industry will be obliged to work with the authorities to develop methods to cover these technical gaps. Given the international nature of many of these issues (*e.g.* many are already part of OECD pesticide initiatives), resources needed to cover this work will be extensive and may take several years to arrive at agreed international standards.

Furthermore, the legal status of the TNGs is also not well established in some Member States. On the one hand certain types of data may not be mandatory yet on the other hand certain Competent Authorities may require data beyond that outlined in the TNGs. The required level of Community harmonisation that the Commission and Industry requires may not be easy to achieve under these conditions.

This burden of high cost (the estimated cost for an active substance dossier is 5–8 MM Euro) against such a background of confusion will stifle innovation. Very few new active substances will be brought to the market in the coming years. This applies equally to alternative methods of microbial control that fall within the directives' definition of a biocidal product *i.e.* '*intended to destroy, deter, render harmless, prevent the action of, or otherwise exert a controlling effect on any harmful organism by chemical or biological means*'. In all cases one must establish the 'mechanism of action' of the proposed novel technology; what causes the observed controlling effect? If the new technology falls within the auspices of the directive then the appropriate supporting data requirements for dossier submission will need to be fulfilled.

7 The Implications Arising from the BPD Risk Assessments Requirements

> *Issue:* The mineral processing industry and the end-users of their products need to agree on; (1) realistic exposure scenario(s) (2) valid emission models (3) field monitoring methods for the application, use and disposal of biocidal products.

Risk estimation is the corner stone of the directive. An assessment of the risk for humans, animals and the environment associated with the application of the biocidal product, the use of treated material and the disposal of treated material/waste is an integral requirement of the authorisation process.

In its simplest form the risk associated with the use of a biocidal product in any particular application area is developed from the following relationships and is essentially a quantitative comparison of exposure concentrations to no-effects concentrations:

RISK *is a function (f)* HAZARD × EXPOSURE
HAZARD = TOXICITY = EFFECT/DOSE
EXPOSURE = DOSE = CONCENTRATION/TIME

The intrinsic hazard or effects assessment is established from the extensive toxicological and eco-toxicological studies performed on the active substance (or the biocidal product when necessary). From this data a dose–response assessment is made and an appropriate no-effect level is derived.

Exposure levels are derived in a three phase process, namely:

1. Description of the life cycle of the biocidal product or active substance(s) in the application.
2. Definition of the exposure scenarios describing 1° and 2° (direct/indirect) human exposure, the various emissions and the 1°/2° environmental compartments exposed.
3. Quantitative estimation of human exposure, environmental release/emissions rates and environmental distribution. This may be achieved through modelling or by field monitoring.

Guidance to date supports the risk assessment principles for general chemical substances already published by the Commission (1996). Consequently, the risk characterisation simply involves a quantitative comparison of the outcome of the hazard/effects assessment with the exposure assessment. For human risk this involves the calculation of the TER (Toxicity : Exposure Ratio) and comparing it with the MOS (Margin Of Safety). For environmental risk the PEC/PNEC ratio (Predicted Environmental Concentration *versus* the Predicted No-Effect Concentration) for the various environmental compartments.

Ideally for biocidal products the guidance documents should:

• Define the exposure scenarios for each product type *i.e.* establish which environmental compartments are exposed as a result of using a biocidal product in any product type.
• Provide guidance on how to develop the most relevant hazard effects (PNEC) data *i.e.* the eco-toxicity testing strategy and methods definition per environmental compartment exposed.
• Provide guidance on how to develop emission/exposure values (PEC) per environmental compartment exposed.
• Define the risk characterisation process.
• Define the decision making process for authorisation.

Through the contributions of the ECB, the eco-toxicology and environmental fate Technical Working Groups, the various initiatives of certain Member States, the Commissions contractors (*e.g.* Haskoning) and reference to the EU technical guidelines for new and existing substances (EURATGD-Part II, 1996) certain progress has been made in most aspects of environmental risk assessment guidance for biocidal products except in the areas highlighted, *i.e.* it is generally recognised that emission scenarios are not defined for many product types and the guidance for estimating the exposure and emission values (PEC) for biocidal products is currently inadequate. Recently a Technical Working Group was created by the Commission entitled EUBEES (EU Biocides Environmental Exposure Scenarios) consisting of Member State, Industry and Institutional experts. Their mission is to:

- Gather information on currently available scenarios and models.
- Define the emission scenarios for product types where no guidance currently exists.
- Propose appropriate predictive models for calculating realistic emission values for product types where no guidance currently exists.

(*N.B. Emission scenarios and predictive models for wood preservatives and anti-foulant product type will not be dealt with by this group but will be discussed within separate/ongoing OECD initiatives.*)

Guidance currently advises that exposure data be obtained by either measuring the environmental concentrations directly or by estimating them using computer-based predictive exposure models such as USES, EUSES, USES II, Mackay fugacity models or CHARM. Unfortunately, both approaches (monitoring and modelling) are sometimes devoid of detail and relevance for certain product types. In the case of monitoring, no guidance is given on; (i) statistical design and relevance of sampling programmes, (ii) validation of trapping and concentration methods, (iii) validation of storage methods and (iv) validation of analytical methods; whilst certain of the predictive models are inappropriate or inadequate for many product types.

Moreover, although various exposure/emission scenarios have been proposed for biocide use relevant to the mineral processing industry there is no agreed industry position on the validity or relevance of these scenarios *e.g.*:

- BIOEXPO, *Release Estimation for 23 biocidal product types*, TNO 1997.
- UK Draft TGD, *Concerning the placing of biocidal products on the market: Common principles and practical procedures for the authorisation and registration of products Version 4*, May 1998.
- *Identification and description of biocide fields of use and the estimation of scores based on the level of human and environmental exposure*, Haskoning 1999.
- The INERIS emission scenario document *'Biocides in paper coating and finishing'*.

Industry is currently faced with the rather unfortunate situation of (1) not knowing exactly which environmental compartments are exposed for certain product types and thus cannot agree on what environmental fate/eco-toxicological testing needs to be done and (2) what tools to use to establish an agreed (between industry/Competent Authority) PEC per product type. It is important to repeat at this point that without the definition of the emission scenarios and an agreement on acceptable predictive emission models the directive will be practically unworkable for environmental risk assessments.

8 The Implications Arising from the BPD Efficacy Requirements

> *Issue:* Industry and the regulatory authorities need to agree on; (1) what constitutes a valid efficacy claim for each product type, (2) the test methods used to generate data in support of these claims and (3) a test method development strategy and plan for the most critical areas where methods are inadequate or missing.

The provision of efficacy data on biocidal products is a core data requirement for all product types. This information takes on central importance within the overall legislation. It not only indicates that a biocidal product can adequately control microorganisms in an application but in defining an efficacy concentration (or range) and method of application it also describes the 'playing field' limits for the development of the associated human and environmental risk assessments.

The BPD defines biocides as *'materials intended to destroy, deter, render harmless, prevent the action of or otherwise exert a controlling effect on harmful organisms.'* Moreover, as indicted within Annex VI (paragraph 89) *'the biocidal product must be shown to give a defined benefit in terms of the level, consistency and duration of control or protection or other intended effects during normal use.'* Efficacy is therefore defined within the directive as the ability of a biocidal product to fulfill the label claims made for it on the proposed label. While there appears to be no official guidance of what constitutes an efficacy claim for the 23 product types the UK draft TGD tentatively listed the parameters that could constitute such a claim. For illustrative purposes the suggested claims are summarised in Table 1 below.

Although this list of options is far from complete it still demonstrates that efficacy claims may often be complex, variable and in certain cases product type specific. It follows therefore that it should be possible to measure efficacy in these terms and that testing methods should provide data that is relevant and pertinent to such a potentially variable efficacy claim. It is the authors' opinion that very few standard laboratory methods are currently capable of fulfilling this objective.

To date no guidelines are available for the efficacy testing of active substances, however, for biocidal products, Paragraph 52 of Annex VI states that testing should be carried out according to Community guidelines if these are available and applicable. Where appropriate other methods can be used *e.g.*

- ISO, CEN or other international methods.
- National standard methods.

Table 1 *Draft efficacy claim*

Product type	PT 1 – 23
Biological spectrum	Bacteria, fungi, yeasts, insects, slime, disinfection
Mode of action	Kill, inhibition of growth, preventative, repellent, growth regulator, knockdown, flushing
Area of use/site of application	As appropriate to product type
Application method	Manual, automatic, brush, spray, dip, impregnation, direct dosing, fog, aerosol, in-can
Duration of effect	???
Expression of application/dose rates	$\%$ w/w, v/v, w/v, $mg\,m^{-2}$, $mg\,m^{-3}$, $kg\,m^{-3}$, $g\,m^{-2}$, dip time

- Industry standard methods.
- Individual producer standards.
- Data from development of the biocidal products.

In fact, there is a recognition that a range of different test data may be required to support label claims. An applicant for authorisation of a biocidal product can confirm their efficacy claim with the provision of data from simple laboratory studies, laboratory simulated-use tests, pilot system studies or field studies according to either standard and/or non-standard test methods.

By allowing this level of flexibility the directive recognises that the laboratory-based 'Standard Test' as a universal tool for predicting accurately and consistently the relative efficacies of all biocidal products within a particular application area is not feasible. It also acknowledges the fact that the available Community standards *may not always* be best fitted to support the Directives' requirements *i.e.* the support of a label claim.

The final TNG 'Guidance on Data Requirements for active substances and biocidal Products, 1999' indicated that test data would only be appropriate if:

- The method(s) adopted measured a response and as appropriate an end-point relevant to the label claim.
- The method(s) should employ chemical/physical/biological conditions relevant to the application.
- The method should include appropriate controls.
- The method should employ a reference product for comparison.
- Dose–response data for dose rates lower than the recommended rate.
- The tests should be based on sound scientific principles. GEP or ISO 9000 is highly recommended.

In 1998, a CEFIC working group composed of industry experts from EPAS (European Producers of Antimicrobial Substances) and EPFP (European Producers of Formulated Preservatives) compiled a listing of over 200 relevant efficacy testing methods and around 100 of these were critically analysed in detail. The following observations of the group are worth highlighting:

* Statistical validation of the methods was rarely available.
* Only a minority of methods were capable of predicting field use-levels.
* Only a minority of methods were capable of ranking biocides.
* When this was possible it was limited to the matrix/samples tested.
* Only a minority of methods defined an internal standard biocide.

In fact, the lack of appropriate efficacy methods is internationally recognised by the authorities and there is an OECD initiative to; (i) improve knowledge on what methods are available for efficacy testing of the different biocidal product types, their validity and any problems they may have and (ii) identify gaps where new methods are needed.

Within the context that an appropriate method is one that fulfills the basic criteria of scientific acceptability and supports a variety of complex label claims. I am sure that the list of gaps is likely to be long. The Commission may have to embark upon a significant program of research, technology development and method validation before 'standard tests' for all applications will be (1) up to the task set for them within the proposed legislation and (2) acceptable to the industries they affect.

9 Conclusions

The BPD will have an increasingly significant impact on the active substance producers, the biocidal product formulators and the end-user industries they serve. The major issues for industry are the following:

Costs: In the early stages the active substance producers will be pre-occupied with the inordinate complexity and costs of dossier preparation. As active substances enter onto Annex 1 the formulators will then start taking on testing costs/industry charges associated with biocidal product authorisation. Competition will be distorted as many SMEs will not be in a position to absorb these costs. Inevitably much of these costs will be passed onto the consumer.

Confusion and Ignorance: In their current state the TNGs are complex, often overlapping, contradictory, incomplete and do not give enough pragmatic advice or guidance. The objective of all producers will be to generate the appropriate amount of data for an acceptable risk assessment and Annex 1 listing in a cost effective manner. At the moment it is not clear how this could be achieved.

Clarification: Producers and formulators will need to work closely with the Competent Authorities to clarify issues of scope, guidance and testing strategy. This will need to occur not only in a general way through the various industry and sector groups but also through one-on-one producer–authority iterative discussion.

Collaboration: To reduce costs certain co-producers and formulators will enter into generic or highly specific data and cost share (task force) agreements. Producers, formulators and end-users will need to work closely together to develop agreed industry input on such issues as efficacy claims, test methods, emission scenarios/models *etc*. This will be resource intensive and time-consuming.

Choice: As the second review regulation is enacted we will see the immediate removal of non-identified active substances and the gradual removal of identified/non-notified active substances (and their formulated biocidal products) from the market. Cost-driven product line rationalisation by the formulators in preparation for authorisation will similarly lead to a reduction in the number of biocidal products. Ultimately there will be less choice for the end-user. Undoubtedly the industries supported by these biocides may be presented with the following potential issues; (1) a decrease in functional protection, (2) hygiene decline, resistance/tolerance, (3) process modifications, (4) end-product re-formulation and (5) higher costs.

Acknowledgments

I would like to thank Bryan Backhouse, Graham Lloyd and Guy Eisenberg for their valuable input to this paper and for their comments on the final text.

References

1. *Opinion of the Scientific Committee on Toxicity, Ecotoxicity and the Environment (CSTEE) on the 'Technical guidance document in support of the Directive 98/8/EC concerning the placing of biocidal products on the market: Guidance on data requirements for active substances and biocidal products – Final version, 7 December 1999'* – Opinion expressed at the 13th CSTEE plenary meeting, Brussels, 4 February 2000.
2. *Comments and Proposals concerning Efficacy Data Requirements in the context of EU Biocides Directive (98/8/EC)*, CEFIC report, 1996/09, pp. 28.

Microbiological Problems in Mineral Slurries – Control Options in 2001

W.G. Guthrie and D.W. Ashworth

BASF MICROCHECK LTD, MERE WAY, RUDDINGTON, NOTTINGHAM, UK

1 Introduction

Mineral slurries are prone to microbial spoilage due to the presence of naturally occurring populations in the source materials (*e.g.* kaolin) and/or through the introduction of contamination during processing. The microorganisms involved can survive and multiply by using trace impurities or dispersant additives as nutrients. The consequences of uncontrolled contamination can be serious with changes in pH, viscosity and depletion of the dispersant leading to separation and caking of the slurry. Additionally, anaerobic sulfate reducing bacteria (SRBs) can grow during storage and shipping and these can cause greying of the slurry due to sulfide by-product formation.

Traditionally, the problems have been addressed by the inclusion of a suitable preservative and in general, the choice of preservative has been fairly wide. However, in recent years, several factors have combined to add to the problems for the slurry manufacturer. Strong pressure from customers in the paper industry has placed cost constraints on the slurry producers with the result that preservative addition levels have been reduced in some cases. Also, increased demand for 'FDA approved' preservatives has reduced the choice available to the slurry manufacturer. These two factors combined have led to increasing incidences of tolerant or resistant bacteria.

In parallel, preservative manufacturers are facing increased scrutiny from regulatory authorities around the world. In Europe, the Biocidal Products Directive (98/8/EC)[1] will add significantly to the cost of supporting active ingredients and this will inevitably lead to a reduction in the number of available preservatives. There is a danger that this reduction in choice may also lead to an increased probability of tolerant/resistant organisms being generated.

This paper will address some of these problem areas and consider options for the mineral slurry producer based on currently available technologies, focusing in particular on Bronopol (Myacide® AS; 2-bromo-2-nitropropane-1,3-diol).

2 Preservation Requirements

2.1 Preservative Demands

Many processes within the mineral slurry industry provide a means of sterilising the resulting suspension – for example the calcining of kaolin or the temperatures produced during the milling of calcium carbonate. However, there are situations where these high temperatures do not apply *e.g.* the production of hydrous kaolin or where contamination is reintroduced during later processing or bulk storage and shipping. Experience has taught that the majority of slurry systems require the addition of one or more preservatives.

When considering the properties of an ideal preservative, the slurry manufacturer might select the following criteria:

1. Broad-spectrum activity
2. Effective over the product shelf-life
3. Safe at use levels
4. Compatible with a wide range of ingredients
5. Water-soluble
6. Effective over a wide pH range
7. Odourless and colourless
8. Low treatment cost

However, it is perhaps a little unrealistic to expect a single active ingredient to meet all of these requirements. Where a shortfall in performance is recognised, then the option to look at combinations of preservatives is always available.

Bearing in mind the problems and constraints referred to earlier, it is worth noting some additional features that are becoming increasingly important:

1. Ability to act synergistically with other biocides
2. Ability to control organisms resistant/tolerant to other biocides
3. Acceptable human and environmental safety profile.

Some examples of how these requirements can be satisfied are presented here.

2.2 Preservative Performance

Contamination problems in mineral slurries are predominantly due to bacteria, more particularly the Gram negative organisms represented by the *Pseudomonas* spp. Anaerobic sulfate reducing bacteria can also be a factor in bulk storage and shipping. Yeasts and fungi may be involved but, in general, these tend to cause fewer problems. It is imperative therefore that the primary preservative has proven efficacy against a broad spectrum of spoilage and anaerobic bacteria.

Bronopol (Myacide AS) is a potent antibacterial agent with some activity against yeasts. It was developed originally by the Boots Company in the UK as a preservative for cosmetics and toiletry products and has been used extensively in these areas since the 1970s. More recently it has been adopted in industrial formulation systems due to its low use concentrations and good compatibility

Table 1 *Minimum Inhibitory Concentrations (MIC) of Bronopol against Gram negative and positive bacteria*

Organism	Number of strains tested	MIC in ppm
Pseudomonas aeruginosa	50	12.5–50
Pseudomonas putida	1	25
Pseudomonas cepacia	1	25
Pseudomonas stutzeri	1	25
Pseudomonas fluorescens	1	25
Escherichia coli	15	12.5–50
Enterobacter aerogenes	1	25
Klebsiella pneumoniae	2	25
Desulphovibrio spp.	9	0.39–12.5
Staphylococcus aureus	30	12.5–30
Staphylococcus epidermidis	2	50
Legionella pneumophila	1	25–50

profile with many commonly used raw materials. The antibacterial activity spectrum is summarised in Table 1.

The potential for Bronopol to act as a preservative in mineral slurries was evaluated both alone and in combination with other preservatives. Where possible, organisms with a claimed resistance or tolerance to commonly used preservatives were used in the screening tests.

3 Methods and Results

3.1 Calcium Carbonate Slurries

A sample of a standard calcium carbonate slurry was received from a large manufacturer in the USA. This sample was subjected to preservative efficacy testing according to the ASTM E 723-91 test protocol.[2] Preservative treated samples were inoculated with a mixed bacterial inoculum containing organisms with a known tolerance or resistance to BIT (1,2-Benzisothiazolin-3-one). Untreated controls were included for reference purposes. The test procedure is outlined below.

3.1.1 Test Samples.
1. Myacide AS (Bronopol) – Batch No. 436786
2. BIT 20% Solution – Batch No. D35
3. Calcium carbonate slurry – ref. Carbital 35

3.1.2 Test Organisms.
1. *Pseudomonas aeruginosa*   BIT resistant isolate
2. *Pseudomonas stutzeri*   BIT resistant isolate
3. *Enterobacter cloacae*   NCTC 10005
4. *Klebsiella aeogenes*   NCTC 418

3.1.3 Test Procedure. The test organisms were subcultured onto Tryptone Soya Agar (TSA) slopes and incubated at 37 °C for 24 hours. After incubation for each inoculum, 6 ml sterile distilled water (SDW) was added to each slope in turn, the organisms washed off and the resultant suspension homogenised. A 1 ml aliquot of the suspension was added to 100 mls SDW and homogenised to form the inoculum.

Known quantities of the mineral slurry were placed in two sterile containers and inoculated with the inoculum. The slurries were thoroughly mixed and 50 ml aliquots were dispensed into sterile, stoppered flasks or capped bottles.

Biocide stock solutions were prepared in sterile containers using SDW. Appropriate volumes of each stock solution were added to give the required concentrations.

Total viable counts (TVCs) were immediately carried out on duplicate controls with no biocide added. The slurries were incubated at 37 °C for 4 hours and TVCs were performed at that time point and at several points over the following 6 weeks. Re-inoculations were carried out after 3 weeks. The resulting bacterial numbers were recorded as colony forming units (cfu)/ml of slurry.

3.1.4 Results and Discussion. The results for samples treated with BIT are shown in Figure 1 and those for the Bronopol treated samples in Figure 2.

The evidence from these tests confirmed that BIT, even at a level of 200 ppm, was unable to control the bacterial contamination based on inoculation and re-inoculation with the resistant/tolerant strains. The results in Figure 2 show that Bronopol was able to control the bacterial contamination at levels down to 50 ppm, even after re-inoculation at the 3-week sample point. The 20 ppm level

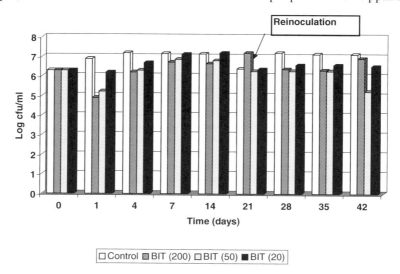

Figure 1 *Preservative efficacy of BIT in a calcium carbonate slurry using the ASTM E723-91 test procedure*

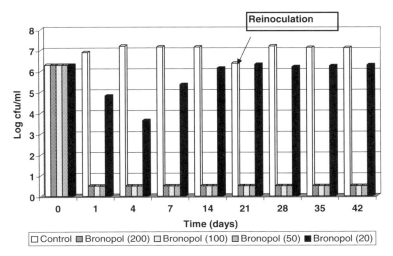

Figure 2 *Preservative efficacy of Bronopol in a calcium carbonate slurry using the ASTM E723-91 test procedure*

showed some initial activity but was unable to sustain control.

These results point to the potential for Bronopol as a preservative in calcium carbonate slurries. In addition, they also indicate the ability of Bronopol to act as a clean-up treatment in slurries already contaminated with organisms showing tolerance or resistance to other preservatives.

3.2 Kaolin Slurries

In a similar way to the testing above, a sample of kaolin slurry was received and subjected to a series of tests based on the ASTM E723-91 procedure. The same BIT resistant inoculum was used but in this instance, the opportunity was taken to examine the activity of chloromethylisothiazolinone/methylisothiazolinone (CIT/MIT) as a preservative both alone and in combination with Bronopol.

3.2.1 Materials.
1. Myacide AS (Bronopol) Batch 436786
2. 1.5% Isothiazolinone (CIT/MIT) solution – Batch unknown
3. Kaolin slurry – ref. Alphagloss slip

3.2.2 Test Organisms.   The inoculum used was identical to that described in Sections *3.1.2/3.1.3* above.

3.2.3 Test Procedure.   The test procedure followed that described in Section 3.1.3 above with the exception that additional tests were carried out on combinations of Bronopol and CIT/MIT. These were designed to assess whether there were any benefits to be obtained over the use of single active ingredients.

3.2.4 Results and Discussion.   The results for the Bronopol concentration series are presented in Figure 3 and show that levels of 50 ppm and above were able to control the contamination during the first three weeks of the test. However, after reinoculation at week 3, none of the treatment levels achieved control.

The results for CIT/MIT are shown in Figure 4 and exhibit a similar pattern to

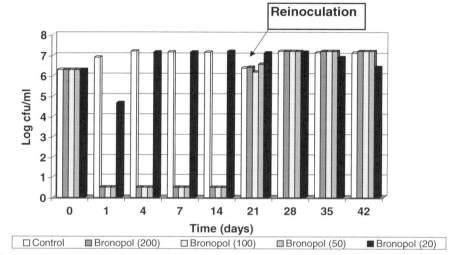

Figure 3 *Preservative performance of Bronopol in a kaolin slurry using the ASTM E723-91 test*

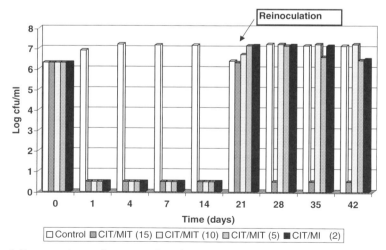

Figure 4 *Preservative performance of Isothiazolinone (CIT/MIT) in a kaolin slurry using the ASTM E723-91 test*

those of Bronopol. For the first three weeks, all levels tested controlled the bacterial contamination. After reinoculation, however, only the highest level (15 ppm) reduced the bacterial levels adequately.

Data generated on a combination of Bronopol and CIT/MIT are presented in Figure 5 along with the relevant results from tests on the single active ingredients. These show that neither 50 ppm Bronopol nor 5 ppm CIT/MIT alone was able to give acceptable performance following reinoculation. However, the combination of 50:5 Bronopol:CIT/MIT achieved complete control throughout the full test duration. These results underline the apparent synergistic benefits of combining these two active ingredients with the result that a more efficient preservative regime is possible.

3.3 Other Applications

The benefits demonstrated from the preservative use of Bronopol alone and in combination with other preservatives have been realised in several other application fields, namely polymer dispersions, adhesives, paper coatings and water-based paints. By way of example, the results of a study in a water-based paint system are reported, where Bronopol is evaluated alone and in combination with 2,2-dibromodicyanobutane (DBDCB).

3.3.1 Test Materials.
1. Myacide AS (Bronopol)
2. DBDCB (min 98%) – Lot No. 1/6/Apr21/7171

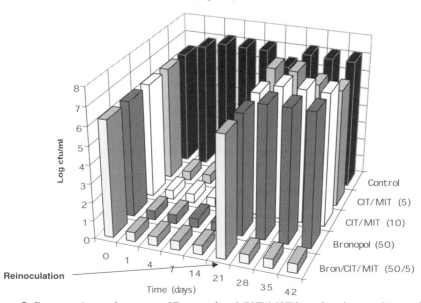

Figure 5 *Preservative performance of Bronopol and CIT/MIT biocides alone and in combination using the ASTM E723-91 test*

3. Low VOC vinyl acetate, ethylene acrylate paint

3.3.2 Test Organisms.
1. *Pseudomonas aeruginosa* ATCC 9027 (NCIMB 8626)
2. *Staphylococcus aureus* NCIMB 9518
3. *Candida albicans* ATCC 10231
4. *Aspergillus niger* ATCC 19404 (IMI 149007)

3.3.3 Test Procedure.
1. 50 ml of the test preparation were placed into 100 ml wide-neck containers, inoculated with 0.5 ml of either the bacterial or fungal mixed inocula and shaken for 5 m at 250 rpm on a rotary shaker. The inoculated test preparation was stored in the dark at $25 \pm 2\,°C$.
2. After 7 days, the test preparation was again shaken for 5 m at 250 rpm. A semi-quantitative determination of colony forming units (cfu) was then carried out by streaking 10 μl onto the surface of a plate of the relevant agar.
3. The TSA plates (bacteria) were incubated at $32 \pm 2\,°C$ and the SAB plates (fungi) were incubated at $25 \pm 2\,°C$ for at least 3 days.
4. The inoculation and cfu determinations were carried out as above until the test preparation either had three consecutive strong growth results from the cfu determination or had six inoculations and cfu determinations.
5. The semi-quantitative determination involved the following scoring system:
    0   No visible growth
    1   Slight growth (individual colonies)
    2   Good growth (colonies merging/confluent growth)
    3   Sample terminated (due to three consecutive 2 results)
The scores recorded over the six challenges are summed and plotted to assess comparative performance of the treatments. High cumulative scores mean poorer performance.

3.3.4 Results and Discussion.   The multiple challenge procedure outlined above represents a relatively severe test for a preservative system. However, systems that perform well under these laboratory conditions should be effective in commercial practice.

Figure 6 shows the results of representative data points from the test series. Here again Bronopol alone gives good bacterial control but poor fungal control at the 100 ppm use level. This result is in line with expectations since Bronopol's spectrum of activity is weak against fungal species. DBDCB at 400 ppm also performs well against bacteria but is not effective against the fungal challenge. The combination of Bronopol:DBDCB (100:400) shows excellent antibacterial and antifungal control – this latter activity being clearly more than a simple additive effect.

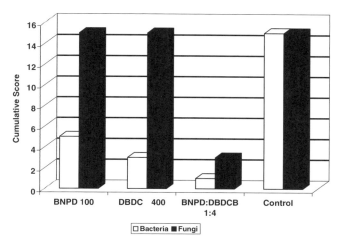

Figure 6 *Preservative efficacy of Bronopol and DBDCB alone and in combination in a water-based paint using a multiple challenge test procedure*

4 Observations and Conclusions

Currently the mineral slurry producer is faced with many difficulties. Commercial pressures are forcing the adoption of cost-cutting strategies that may influence the choice of preservative regime. At the same time, however, there is the continued need to satisfy human and environmental safety requirements. Preservation may seem to be a minor consideration but neglecting it can lead to costly problems. The appearance of tolerant or resistant organisms is evidence of the type of problem that arises where cost pressures have resulted in a reduction in preservative levels. This fact underlines the importance of selecting single or combination preservatives that have a high level of antibacterial activity coupled with the ability to deal with intransigent contamination.

The results reported here indicate the potential for Bronopol (Myacide AS) as a preservative in mineral slurry systems. In particular, its ability to control organisms that are tolerant or resistant to other biocide chemistries is a valuable feature. Sondossi *et al.*[3] have reported on the activity of Bronopol against organisms that are resistant to formaldehyde. The results presented here indicate that this ability also covers BIT resistant organisms. The explanation for this may be related to some unique facet of Bronopol's mode of action and this fact is supported by published work.[4,5] Overall, it suggests that Bronopol could be used in clean-up regimes to recover contaminated product.

Another important property is Bronopol's ability to act synergistically with other preservatives. The data presented here confirm the benefits of combining Bronopol with isothiazolinones (CIT/MIT) in mineral slurries and points to synergistic activity in combination with 2,2-dibromodicyanobutane (DBDCB) in a water-based paint system. The careful use of combinations of this sort could allow the slurry producer to optimise the preservative with resulting cost benefits.

Acknowledgements

The authors wish to acknowledge the support of Cath Lewis and Lawrence Staniforth in performing the microbiological efficacy tests reported in this paper.

References

1. *Official Journal of the European Communities*, 1998, **41**, L123.
2. *Annual Book of ASTM Standards*, American Society for Testing and Materials, 1995, Section 11, **11.05**.
3. M. Sondossi, H.W. Rossmore and J.W. Wireman, *J. Indust Microbiol.*, 1986, **1**, 87–96.
4. P. Higdon, PhD Thesis, 1985, North East Surrey College of Technology, Ewell, Surrey, England.
5. J.S. Chapman, M.A. Diehl and R.C. Lyman, *Journal of Industrial Microbiology*, 1993, **12**, 403–407.

® Myacide is a registered trademark of BASF AG.

Modern Microbiological Problems and Solutions for the Pigment Slurry Industry

J.L. Martin

ONDEO NALCO, ONE NALCO CENTER, NAPERVILLE, IL 60563, USA

1 Introduction

The water borne coating market is growing steadily. While concerns about safety and the environment help drive growth in the market, escalating consumer demands for better quality coatings and ever improving manufacturing technology continue to drive improvements in this industry. In fact, today's commodity coatings are the 'high grade coatings' of 10 years ago. In order to guarantee consistent quality in such a demanding environment, manufacturers precisely control every aspect in the formulation of a product.

Pigment slurries are one of the major components in paints and coatings that contribute to or affect the performance of a formulation. The nature of pigment slurries makes them extremely susceptible to microbiological contamination that can degrade the final formulation. Often biodeterioration can be prevented with plant sanitation and the use of a preservative, but formulators must not rely on the precept that a particular preservative that works today will always remain effective down the road. In complex environments, contaminants can enter the system at different points, or a formulation change may render the original preservative less effective. Manufacturers must remain aware of trends in the industry to ensure their products are properly protected.

A new liquid preservative has been designed specifically to protect pigment slurries for use in paints and coatings. The emergence of a new problem bacterium, *Methylobacterium* sp., led to an investigation of new combinations of FDA-approved biocides that could provide cost-effective protection against this difficult-to-control microorganism. The new biocide must offer both short-term and long-term protection over a broad pH range and be compatible with other components in a paint or coating formulation. A combination was found that offers both safety and effective protection that cannot be matched by existing biocides.

1.1 Damaging Effects of Microorganisms in Paints and Coatings

The chemical and/or physical alteration of a product caused by microbiological degradation is a major problem for many coating industries. Failure to consider the effects microorganisms have on the formulation can lead to major problems. Microorganisms degrade a paint or coating by utilising the raw materials (thickeners, defoamers, colloids, surfactants, emulsions, pigments, *etc.*) in the formulation as their food source. To appreciate the significance of microorganisms, we should think of these organisms as minute factories capable of carrying out complex chemical conversions that alter almost any property of a coating, including a change in the rheological properties, pH, colour and odour of the system.

A loss in viscosity occurs when enzymes, organic catalysts manufactured by bacteria and/or fungi, start to destroy the thickener. These enzymes are capable of triggering reactions quickly and for indefinite periods. One enzyme molecule can effectively change millions of raw material molecules to undesirable end products. The result can be a paint or coating that loses its elastic and elongational properties.

Organic acids produced by the organisms can cause a drop in the pH of a formulation, resulting in physical instability. These same organic acids, along with other metabolic byproducts of the microorganisms, can also impart unwanted odours and colours which alter the appearance and absorption of the paint. Incorporation of a preservative is recommended to prevent these problems from occurring.

1.2 Potential Areas of Contamination

Problem microorganisms can be introduced into a paint at almost any point of the manufacturing process. It is essential to identify the areas where contamination can occur to develop an effective preservation package. Over the years, the most problematic areas in many manufacturing facilities have been identified by conducting plant audits. From these audits, a list of recommendations for keeping certain equipment and materials uncontaminated has been developed. The most important areas to consider when assessing plant hygiene are presented below in Table 1.

Identifying and controlling problem spots in a manufacturing process can limit the extent of contamination. However, only the use of a preservative, coupled with good plant hygiene, can prevent microbiological outbreaks and maintain the integrity of the formulation through the plant facility, storage, transfer and transport to the consumer.

1.3 Evaluation of Preservatives

Formulators faced with the job of incorporating a biocide into a paint or coating to overcome biodeterioration have a wide variety of chemical agents to choose from and encounter several different factors that influence the performance of a

Table 1 *Recommendations for effective plant hygiene*

Production area	Recommendations

Raw materials:
  Pigment slurries
  Thickeners
  Surfactants
  Defoamers
  Colloids
  Water
	A. Periodically check the microbiological condition of the raw materials. Cover the materials when not in use. Uncovered raw materials are susceptible to microbial growth and degradation.
	B. Periodically change measuring pails. Over time, old pails will accumulate excess raw materials that can become susceptible to microbial attack.

Equipment:
  Storage tanks
  Mixing tanks
  Filling trays
  Pipes
  Rail cars
  Pumps
  Hoses
	A. Remove skins from mixing and storage tanks. Periodically scrape tanks to bare metal and either steam clean or rinse with a disinfectant.
	B. Remove residual water from filling trays, pipes, tanks, pumps, *etc.* Water left in the equipment will support microbial growth and can serve as an inoculum for fresh materials. If water cannot be drained, treat with an effective biocide.
	C. Properly store equipment when not in use. Hoses should be stored on racks in a vertical position so that any residual product drains out.

Personnel	A. Educate personnel to look for signs of microbial contamination:
	    Colour changes
	    Viscosity drops
	    pH drop
	    Odours
	    Gassing

preservative when incorporated into a specific application. Understanding the effects of the microorganisms on a system, the manufacturing process, product chemistry and contamination sites, allows a qualified biocide supplier to recommend the most cost-effective treatment programme. Recommendations are usually made after conducting extensive microbiological studies. A properly designed microbiological study will consider the types of microorganisms, the pH and temperature at which biocide addition occurs and whether there are any chemicals (oxidisers and/or reducing agents) present in the formulation that may interfere with the preservation properties of a particular biocide. All of these factors are critical to the success of the biocide programme.

Historically, a large number of bacterial species have been isolated from pigment slurries. The most common organisms have been identified as *Pseudomonas aeruginosa, Pseudomonas* sp., *Enterobacter* sp. and *Alcaligenes faecalis.* Recently, a pigmented organism has emerged as a problem in slurries. This organism has been identified as *Methylobacterium* sp. and, like other common slurry contaminants, is found in the air, soil and water. This organism uses carbon from the gasses produced by other bacteria contaminating the system. This organism then becomes the predominant species and severely alters the appearance of the slurry. Strains of the *Methylobacterium* may be pigmented (yellow, ochre, pink or red) and have been known to turn white slurries pink or

Table 2 *Biocides tested and percent active*

Biocides	Percent active
1,2-dibromo-2,4-dicyanobutane (12% active) + 2-bromo-2-nitropropane-1,3-diol (6% active)	18% total active
2-bromo-2-nitropropane-1,3-diol	98% active
1,2-benzisothiazolin-3-one	19% active
5-chloro-2-methyl-4-isothiazolin-3-one and 2-methyl-4-isothiazolin-3-one	1.5% total active
1,5-pentanedial	50% active
1,2-dibromo-2,4-dicyanobutane	25% active

red. This organism has also shown resistance to most preservatives in use today.

A new liquid preservative blend has been developed to prevent microbiological degradation of pigment slurries as well as the paints and coatings that utilise the pigment. The preservative, 1,2-dibromo-2,4-dicyanobutane and 2-bromo-2-nitropropane-1,3-diol in dipropylene glycol solvent, was evaluated in various pigment slurries and its performance was tested against several other industrial biocides. For the preservation study, various concentrations of each biocide were added to the test system. The chemical name of each biocide, and its percent active, are listed below in Table 2.

2 Materials and Methods

All the slurries were heavily contaminated with bacteria upon receipt and tested following the Return to Sterility Preservation Test Method (see Appendix). The pigment slurries (listed below in Table 3) were obtained from various domestic and international suppliers. The level of biocide needed to return the system to an uncontaminated condition was designated as providing short-term protection (<3 months). The level of biocide needed to protect the system from an inoculation of the systems own contaminants was designated as providing long-term protection (6 months or greater).

The test organisms (listed below in Table 4) utilised for the study were wild strain bacteria isolated from the slurries.

The organisms were identified through the gas chromatographical analysis of the fatty acids in their cell walls. The general procedure involved culturing of the bacteria and collecting the cell mass, followed by saponification of the cells with sodium hydroxide in methanol. The liberated fatty acids were converted to the methyl esters with hydrochloric acid in methanol. The methyl esters were extracted with hexane/methyl-*tert* butyl ether and analysed by gas chromatography (GC). The GC analysis produces a chromatogram demonstrating a distinct pattern for each group or species of bacteria. The fatty acids that are present in bacterial cells vary in chain length consisting of long chain fatty acids.

Table 3 *Pigment slurries*

Sample identification	Supplier location	pH
Slurry 1	USA	8.6
Slurry 2	USA	8.9
Slurry 3	USA	9.0
Slurry 4	Sweden	8.3
Slurry 5	Italy	8.6
Slurry 6	Taiwan	9.2
Slurry 7	Australia	9.5
Slurry 8	Japan	9.5

Table 4 *Test microorganisms*

Pseudomonas sp.	Aerobic, Gram negative rods, widely distributed in nature
Pseudomonas aeruginosa	Aerobic, Gram negative rods, widely distributed in nature
Enterobacter sp.	Facultatively anaerobic, Gram negative rods, widely distributed in nature
Alcaligenes faecalis	Obligately aerobic, Gram negative rods or cocci, occur in water or soil
Methylobacterium sp.	Utilise methane as carbon source, occur in air, soil, water or leaves

Table 5 *Top three effective products (ppm active)*

	pH	1,2-Dibromo-2,4-dicyanobutane +2-Bromo-1,3-diol	2-Bromo-2-nitropropane-1,3-diol	1,5-Pentanedial
Slurry 1	8.6	45 ppm	49 ppm	125 ppm
Slurry 2	8.9	90 ppm	49 ppm	125 ppm
Slurry 3	9.0	45 ppm	73.5 ppm	125 ppm
Slurry 4	8.3	13.5 ppm	14.7 ppm	25 ppm
Slurry 5	8.6	13.5 ppm	9.8 ppm	250 ppm
Slurry 6	9.2	27 ppm	9.8 ppm	25 ppm
Slurry 7	9.5	13.5 ppm	9.8 ppm	25 ppm
Slurry 8	9.5	180 ppm	490 ppm	250 ppm

3 Results

The top three effective products in each system providing short-term and long-term protection are listed in Table 5.

Slurries #1, #3, #4 and #8 show the new liquid blend, 1,2-dibromo-2,4-dicyanobutane + 2-bromo-2-nitropropane-1,3-diol, as being the most effective product, while the 2-bromo-2-nitropropane-1,3-diol was most effective in slurries #2, #5, #6 and #7. In order to appreciate the advantages of this new blend over the products tested, the observations from the study of each of the products are listed in the discussion.

4 Discussion

This study was designed to evaluate the performance of commonly used biocides against the performance of a new liquid blend. In order to choose an effective preservative package, the two main criteria the biocide must have, are (i) to be approved by the US Food and Drug Administration (FDA) under sections 21 CFR 176.170 and 21 CFR 176.180 which cover components of paper and paperboard in contact with foods and (ii) to be designated as a safe biocide (low toxicity, non-sensitiser, easy to handle). The three most commonly used products; (i) 5-chloro-2-methyl-4-isothiazolin-3-one + 2-methyl-4-isothiazolin-3-one, (ii) 1,5-pentanedial and (iii) 1,2-benzisothiazolin-3-one, currently meet the above criteria. The new liquid blend, 1,2-dibromo-2,4-dicyanobutane + 2-bromo-2-nitropropane-1,3-diol was also designed to meet the criteria.

4.1 1,2-Dibromo-2,4-dicyanobutane + 2-Bromo-2-nitropropane-1,3-diol

This product was among the top three effective performers in all systems. This product's main advantage over the others tested is that it effectively controlled the *Methylobacterium* sp. in all slurry samples. This is also a combination biocide that provides beneficial synergy and enables the use of lower dosages. The result is a safe biocide that decreases exposure to the handler and the end-user. The product is FDA-approved under the specified clearances. It is very effective over a pH range of 8.3–9.5 and unlike the 1,5-pentanedial, it produced no offensive odour.

4.2 2-Bromo-2-nitropropane-1,3-diol

This product is very effective. This product was among the top three effective products providing excellent long-term protection in all slurries. However, one problem is that, at dosages between 245–980 ppm, the product can cause an increase in the viscosity of some of the slurries. Also, this product only has FDA approval under 21CFR 176.170 for 100 ppm active, based on the solids of the slurry. Thus, in some cases this could pose a problem if more than 100 ppm active is needed to provide adequate protection to a slurry. This molecule has also been shown to be unstable over a long period of time at a pH of 8.0 and greater. It can rapidly degrade in the presence of some oxidising and reducing agents. The instability of the product could make the slurry more susceptible to microbial growth and degradation. Routine testing with the use of this molecule is highly recommended.

4.3 1,2-Benzisothiazolin-3-one

This product did not perform well in any of the slurries. It is very expensive and requires very high dosages to provide adequate protection to a system. This molecule was not effective in controlling the *Methylobacterium* sp.

4.4 5-Chloro-2-methyl-4-isothiazolin-3-one and 2-methyl-4-isothiazolin-3-one

Historically, this molecule has been very effective in controlling microbial growth and degradation in pigmented slurries. However, its performance against the *Methylobacterium* sp. was poor.

4.5 1,5-Pentanedial

This product was among the top three performers, providing excellent long-term protection to all slurries. Like the 5-chloro-2-methyl-4-isothiazolin-3-one and 2-methyl-4-isothiazolin-3-one, this molecule has been very effective in controlling microbial growth and degradation in pigmented slurries and worked well against the *Methylobacterium* sp. The major problem with this biocide in pigment slurries is its odour, which is very pungent. An effort to use other products for slurry preservation is now underway both domestically and internationally.

4.6 1,2-Dibromo-2,4-dicyanobutane

This product performed well but was not as cost effective as the other products.

5 Conclusions

1. The new liquid blend, 1,2-dibromo-2,4-dicyanobutane + 2-bromo-2-nitropropane-1,3-diol, was very effective in controlling the *Methylobacterium* sp. It is the recommended product of choice for the preservation of pigmented slurries.
2. The 2-bromo-2-nitropropane-1,3-diol was effective in controlling the *Methylobacterium* sp. but, due its stability issues in pigment slurries, it is recommended as the second product of choice.
3. The 1,5-pentanedial was effective in controlling the *Methylobacterium* sp. This product's pungent odour makes it an unlikely candidate for the preservation of pigment slurries.
4. The sulfur-based non-oxidising biocides were consistently ineffective in this study.

The development of new biocides is very expensive and bothersome. Yet, due to strict government regulations, consumer consciousness and concerns about the environment, manufacturers are under pressure to decrease the use of biocides in their formulations. The goal is to develop 'safer biocides'. To meet this objective, manufacturers are demanding more potent combination biocides based on existing chemistries. No single preservative can solve every microbiological problem. As new bacteria emerge and formulations change, the only way to stay on top of changing preservation needs is by working with a qualified biocide supplier whose goal is to develop safer chemical agents. Today, the development of 1,2-dibromo-2,4-dicyanobutane + 2-bromo-2-nitropropane-1,3-diol is the recommended product of choice for effective preservation of pigment slurries.

Appendix
Return to Sterility Preservation Test

Purpose. To determine an effective biocide treatment level to recover a contaminated aqueous system.

Procedure. Individual samples are weighed out from the contaminated test material. Preservatives are then added to the samples in a ladder series of levels (*e.g.* 500, 1000, 1500, 2000 ppm) and mixed thoroughly. The samples are then incubated at 28–30 °C, 85% relative humidity and tested for microbial growth by streaking on the appropriate agar media at intervals of 48 hours and seven days after biocide addition. Following the initial 7 day test period, the microorganisms isolated from the contaminated system are then prepared in Sterile Butterfield Buffer and used as the inoculum. The samples are then inoculated with 0.25 ml of inoculum per 25 g of test material. This provides an inoculum of approximately 10^6 microorganisms per gram of aqueous system. The samples are then incubated at 28–30 °C, 85% relative humidity and tested for microbial growth by streaking on the appropriate agar media at intervals of 48 hours and 7 days following inoculation.

Interpretation. The effective biocide treatment is the lowest concentration that produced sterility in the system within 7 days of biocide addition and inoculation.

   a. Short-term protection – a sample that returns to sterility within 7 days of biocide addition is said to be adequately protected for short-term storage (less than 3 months).
   b. Long-term protection – a sample that returns to sterility within 7 days of both biocide addition and inoculation is said to be adequately protected for long-term storage (greater than 6 months).

References

1. A. Eilender and R. Oppermann, 'Biocides: Bactericides and Fungicides' in *Handbook of Coating Additives*, 1986, pp. 177–193.
2. J. Holt, K. Noel, P. Sneath, J. Staley and S. Williams, *Bergey's Manual of Determinative Bacteriology – 9*, 1994, pp. 75, 78–80, 93–94, 178.
3. V. King, 'Bactericides, Fungicides and Algaecides' in *Paint and Coating Testing Manual*, 1995, pp. 261–267.
4. A.D. Russel, W.B. Hugo and G.A.J. Ayliffe, 'Preservation of Specialized Areas' in *Principles and Practice of Disinfection, Preservation and Sterilization*, 1992, pp. 431–435.
5. *Identification of Bacteria by Analysis of Cellular Fatty Acids, Bulletin 767C*. Supelco Inc., 1977.
6. G. Weber, 'How to Establish and Maintain Effective Plant Sanitation' in *The Adhesives and Sealants Agenda Europe*, 1998, pp. 88.

Subject Index

s/o m.

Books are to be returned on or before
the last date below.

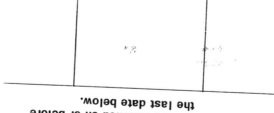